Advanced Sciences and Technologies for Security Applications

Indexed by SCOPUS

The series Advanced Sciences and Technologies for Security Applications comprises interdisciplinary research covering the theory, foundations and domain-specific topics pertaining to security. Publications within the series are peer-reviewed monographs and edited works in the areas of:

- biological and chemical threat recognition and detection (e.g., biosensors, aerosols, forensics)
- crisis and disaster management
- terrorism
- cyber security and secure information systems (e.g., encryption, optical and photonic systems)
- traditional and non-traditional security
- energy, food and resource security
- economic security and securitization (including associated infrastructures)
- transnational crime
- human security and health security
- social, political and psychological aspects of security
- recognition and identification (e.g., optical imaging, biometrics, authentication and verification)
- smart surveillance systems
- applications of theoretical frameworks and methodologies (e.g., grounded theory, complexity, network sciences, modelling and simulation)

Together, the high-quality contributions to this series provide a cross-disciplinary overview of forefront research endeavours aiming to make the world a safer place.

The editors encourage prospective authors to correspond with them in advance of submitting a manuscript. Submission of manuscripts should be made to the Editor-in-Chief or one of the Editors.

Munir Zamir · Christian Kaunert

Strategic Communications' Role in Counter-Terrorism

The Power of Counter-Narratives

Springer

Munir Zamir
Faculty of Life Sciences and Education
International Centre for Policing
and Security
University of South Wales
Pontypridd, UK

Christian Kaunert
Faculty of Life Sciences and Education
International Centre for Policing
and Security
University of South Wales
Pontypridd, UK

School of Law and Government
DCU Conflict Institute
Dublin City University
Dublin, Ireland

ISSN 1613-5113 ISSN 2363-9466 (electronic)
Advanced Sciences and Technologies for Security Applications
ISBN 978-3-031-89109-0 ISBN 978-3-031-89110-6 (eBook)
https://doi.org/10.1007/978-3-031-89110-6

This work received financial support from the Czech Science Foundation under standard grant no. 24-10160S.

This Springer imprint is published by the registered company Springer Nature Switzerland AG
The registered company address is: Gewerbestrasse 11, 6330 Cham, Switzerland

This book is dedicated to the loving memory of my late father—Mr. Najmul Hassan Zamir, to my wife Saima, my son Ibrahim and daughter Aminah.

—Dr. Munir Zamir

To my daughter Elisabeth and my son Raphael. You are my eternal pride until my last breath. To my wife Sarah—I will always love you. To my parents—many thanks for all your support and love.

—Prof. Christian Kaunert

Foreword

In today's rapidly changing and to some extent, evolving security environment, the strategic communications landscape demands a paradigm shift. The authors of this ground-breaking research, Dr. Munir Zamir and Prof. Christian Kaunert present a timely and critical analysis of how counter-extremism efforts must adapt to the complexities of the digital age. Their work explores a vital truth: strategic communications cannot rely on outdated, static and inappropriate approaches to counter extremist narratives. Instead, they advocate for an approach rooted in understanding human agency, socialization, and context-specific responses—principles that are often overlooked and ignored in traditional counter-terrorism strategies.

The authors adopt a critical approach and highlight the limitations of current frameworks, particularly the overreliance on short-term metrics and psychological models that fail to account for the broader social dynamics at play. Counter-narratives are often evaluated based on their digital reach rather than their real impact on behaviour. By dissecting initiatives such as Meta's "Counter Conversations," Google's "Search Redirect," and the Abdullah-X campaign, they reveal a key insight: success depends on the intelligent design of narratives that resonate authentically within diverse communities, creating lasting engagement and behavioural shifts.

This work introduces the Socialisation and Reasoned Action Strategic Communication (SoRaSCo) model, a comprehensive framework designed to address the gap between theoretical ambition and practical application. The authors demonstrate how integrating socialization theory with strategic communications offers a nuanced understanding of how individuals engage with messages in both digital and physical environments. By emphasizing the iterative nature of effective communication—where feedback loops and cultural context drive message adaptation—the model provides a pathway for developing campaigns that are targeted, sustainable and impactful.

At the heart of this research is the idea that countering extremist ideologies is not simply about rejecting harmful narratives but about offering compelling alternatives that foster inclusion and belonging. Extremist recruitment often thrives on feelings of exclusion, alienation, and mistrust—factors that cannot be addressed by one-size-fits-all messaging. The authors argue that strategic communications must recognize the diversity within target audiences and craft interventions that reflect this complexity.

A major contribution of this work lies in its critique of the prevailing focus on so-called "vanity metrics." The authors caution against equating digital engagement with success and instead advocate for robust evaluation frameworks that assess long-term impact. This shift in focus requires collaboration across sectors, including policymakers, practitioners, and researchers, to ensure that interventions are not only effective but ethically grounded.

The digital landscape continues to evolve, with new platforms, algorithms, and audience behaviours reshaping how information is consumed and acted upon. The authors challenge stakeholders to remain adaptive, leveraging technology while maintaining a human-centred approach. Their work underscores that strategic communications must move beyond reactionary tactics to proactive strategies that anticipate and mitigate emerging threats.

This work is a call to action for those involved in national security, communications, and policy development. It highlights the need for continuous learning and adaptation, reminding us that effective counter-narratives are not static but living, evolving mechanisms. By blending theoretical rigor with practical insights, this research offers a blueprint for designing strategic communications that are as resilient and dynamic as the challenges they seek to address.

The authors' vision is not just about countering extremism but about fostering societal cohesion through communication strategies that inspire trust and resilience. In doing so, they provide a foundation for future efforts to safeguard security and stability in an increasingly interconnected world. I strongly recommend this reading and the authors to all of those involved in strategic communications and counter extremism.

Simon Bergman
CEO, M+C Saatchi World Services
London, UK

Preface

Extremism is everywhere. It can be found in all walks of life and often in the most unlikely contexts. The internet, however, is not one of these unlikely settings. Online extremism, hate speech and disinformation are truly 'at home' on social media platforms. Technology companies are constantly grappling with the need to safeguard online communities from harm. One such effort over the past decade has been to design and disseminate online counter-extremism content, known as 'counternarratives', in an attempt to reduce the allure of extremist ideology, narratives and recruitment. This involves the use of strategic communication as a tool to reach and engage audiences with a view to influencing behavioural shifts. Tech giants such as Meta and Google have found themselves caught between government wishes for censorship and 'take-downs' of extremist content, online user needs for more and more unregulated content, and a corporate drive to be profitable whilst also seeming to take social moral responsibility seriously. Also in this milieu are community sector organisations, which have grassroots understanding and elements of cultural knowhow, proving vital to counter-extremism efforts.

This book offers policy makers, practitioners and researchers a detailed look at the conceptual, operational and evaluative elements that make-up counternarrative efforts through strategically aligned communications objectives. It explores the 'what' why and how of counternarrative work, using detailed discourse analysis of three groundbreaking case studies. Each case study is different in its tactics and methods, but all three were designed with the overarching aim of positively impacting the need to challenge extremist content online. The research contained covers critical analysis of terrorism and radicalisation discourses, offers a new framework for creating and measuring counternarrative efforts, and assess the crucial question; what makes a communications effort 'successful'?

Pontypridd, UK

Munir Zamir
Christian Kaunert

Acknowledgements

Dr. Munir Zamir is grateful to his wife and children for their love and support, Prof. Christian Kaunert for providing him with the opportunity to extend his learning throughout his Ph.D., and Mr. Simon Bergman, Mr. Nathaniel Hill and Mr. Paul Smyth from M&C Saatchi World Services for their unwavering support. He would also like to extend his gratitude to Dr. Erin Saltman from the Global Forum to Counter Terrorism (GIFCT), for always being a willing advocate in his work, Mr. Rashad Ali and Ms. Katherine Szamuely for efforts that go far beyond the norm in fellowship.

This book is a qualitative case study involving detailed discourse analysis via case study research into communications efforts to counter extremism and terrorism online. Some of the most innovative and trailblazing counter extremism efforts have been reviewed, analysed and critical examined, in order to create a robust framework for understanding what constituted effective counter narrative efforts in this domain of work.

This book was inspired by Dr. Munir Zamir's research at the University of the South Wales. For this, he is grateful to his mentors who read and critiqued the original research, which provided the foundation for this book. Thank you, Prof. Sarah Leonard, Katherine Szamuely, and Rashad Ali for being vital supporters and critical friends of the research and writing process.

The findings presented in this book are based on research supported by various institutions, including Dublin City University, the University of South Wales, Antwerp Management School and the Institute Barcelona d'Estudis Internacionals (IBEI). Professor Christian Kaunert also wishes to thank the European Union for funding the Marie Curie Doctoral Training Network EU-GLOCTER (coordinated by Dublin City University). This book has also been written with the support of the Erasmus+ programme of the European Union (Jean Monnet Actions), notably Prof. Christian Kaunert's Jean Monnet Chair EUCTIRL in EU Counter-Terrorism, as well the Jean Monnet Centre of Excellence VIRTUE, the Jean Monnet Module EUHYBRID, and, finally, the Jean Monnet Network in EU Counter-Terrorism EUCTER. While writing this book, we also benefited enormously from the comments and insights of numerous experts in the field. We thank them for their time and support.

We are grateful to the entire team at Springer nature, for their support, encouragement, and understanding throughout the publication process. We are deeply grateful for the editorial support and expertise put in to get this book to the readers.

We reserve a final mention for those individuals who have offered tremendous support throughout this book adventure. First and foremost, we sincerely thank our respective parents and families, who have always been a source of inspiration. We also express our appreciation to our friends, who have provided encouragement at just the right moment.

Contents

1 Introduction 1
Originality and Significance 3
State of the Art 4
Theoretical Framework and Approach 5
Summary of the Methodological Approach 7
References 9

2 Radicalisation and Counter-Terrorism—Reviewing the Problems in Strategic Communications Role in Counter-Terrorism 11
Socialisation Theory 12
Anthony Giddens: Structuration Approach 16
Pierre Bourdieu: Social Agency and 'Habitus' 17
Critical Theory: Zygmunt Bauman and Modernity as a Key Social 'Agent' 17
Post-Modernism and Socialisation 18
Social Change and Political Identity—Francis Fukuyama 19
Socialisation as a Segue into Radicalisation Discourse 20
Radicalisation and Deradicalisation Discourses 21
Radicalisation: A Confusing or Contested Concept? 22
Lack of Conceptual Clarity and Disparity of Radicalisation and P/CVE Contexts 24
Radicalisation and Psychology-Based Concepts Within P/CVE 25
Radicalisation and Pathway Process-Based Approaches 27
Issues with Radicalisation Terminology and Usage in P/CVE 30
Deradicalisation 31
Outlier Discourses Within Deradicalisation Studies 32
Deradicalisation and a Perceived Lack of Clarity 33
Strategic Communications Studies 35
Strategic Communications as an Extension of Communication 37
The Development of Communications Studies 39

Symbolism and Meanings in Communications 39
The Reasoned Action Model (RAM) of Behaviour Change 41
P/CVE Perspectives and Gaps in Understanding of the Sector 48
Application of Concepts Within P/CVE Efforts 49
Summary of Socialisation, Communicative Behaviour Change Models and P/CVE Sector 55
Conclusion 57
References 58

3 The Socialisation and Reasoned Action Strategic Communications' Model 65
Introduction 65
Benefits of Adopting a Socialisation-Based Communications Theory Framework 67
Criticisms of Socialisation-Based Approaches 68
Developing and Improving the Socialisation and Communication Model Within P/CVE 69
SoRaSCo—'Socialisation and Reasoned Action Strategic Communications' Model 70
Significance and Originality of the Research 72
Research Design and Method 76
Analysis Criteria—RAM, Socialisation Theory and Communications Approaches/Contexts 80
RAM and Behavioural Intention 81
RAM and Attitudes Towards Behaviour 81
RAM and Subjective Norms 82
RAM and Perceived Behavioural Control 82
RAM and Environmental/Structural/Background Contextual Factors 83
RAM, Feedback and Iterative Approaches 83
References 84

4 Abdullah X Project 87
Introduction 88
The Origins of AX 88
Conceptual Design and Broad Aims of the AX Project 89
Discourse Analysis of the AX Project 90
Framework and Criteria Analysis: RAM, Socialisation, and Communication Models 93
Transcript of Abdullah-X's 'Five Considerations for a Muslim on Syria' Video 97
RAM in Conjunction with Strategic Communications Analysis 97

Analysis of AX Example 2 Video's Narrative 100
Strategic Communications and Dissemination Considerations 117
Campaign Objectives and Outcomes 125
Analysis of AX Using the SoRaSCo Integrative Model 128
Broader Topics of Interest to the AX Case Study 130
References 133

5 Meta's Counter Conversations Project and Google's Search Redirect Project 135
Introduction 136
Background to the Campaigns 137
Case Study 2: The Counter Conversations Project (CC) 137
Design Considerations 138
Socialisation Theory Considerations Within the Project's Parameters 142
CC Project Outcomes Applied to the RAM 142
Analysis of CC Using SoRaSCo Model 147
RAM Approach and Socialisation Analysis of CC 147
Strategic Communications Perspectives as Applied to CC 148
Learnings and Further Considerations of the CC Project 149
Case Study 3: Search Redirect Project (SR) 150
Background Context to SR 150
SR's Objectives 151
Design and Delivery Issues for SR 152
RAM Approach and Socialisation Analysis of SR 154
Strategic Communications Perspectives as Applied to SR 155
Key Learnings from the Analysis of SR 155
References 156

6 Broader Strat Comms and P/CVE Reflections 159
Introduction 159
Understanding and Contextualising Online P/CVE Communications Efforts 160
Design and Delivery of Campaigns: Tools and Tactics of Interventions 160
Engagement, Measurement, and Evaluation: Behaviour Change and Sentiment Analysis 160
Understanding and Contextualising Online P/CVE Communications Efforts 161
Design and Delivery of Campaigns: Tools and Tactics of Interventions 161
Engagement, Measurement and Evaluation: Behaviour Change and Sentiment Analysis 166
Summary of Chapter Findings and Key Learnings 173
References 174

7 Conclusion 177
Original Contribution and Significance 178
General and Specific Results of the Case Studies 180
Limitations 182
Challenges to Existing Perceptions from the Research Findings 183
Research Limitations and How These Findings Can Be Built Upon 186
References 190

About the Authors

Dr. Munir Zamir is a Visiting Fellow of the University of South Wales (International Centre for Policing and Security), Strategic Engagement Director for M+C Saatchi World Services–Middle East, and Independent Advisory Committee Member (IAC) to the Global Internet Forum to Counter Terrorism (GIFCT).

Prof. Christian Kaunert is Professor of International Security at Dublin City University, Ireland, Professor of Policing and Security and Director of the International Centre for Policing and Security, and Jean Monnet Director (Chair, Centre, and Network) on EU Counter-terrorism at the University of South Wales.

Glossary of Terms and Abbreviations

Glossary

Algorithm	A tool used in social media to deliver pieces of content to users based on assumed likes and preferences
Behaviour Change	An attempt to modify or alter existing or possible future behaviour through some form of intervention via an external source
Tech	Reference to the technology sector and primarily used as a catch all term to denote both the sector, its main protagonists and their mode of business

Abbreviations

AVE	Against Violent Extremism Network
AX	Abdullah-X project/campaign
CC	Counter Conversations project (also known as One2One)
CN	Counternarrative
CS	Counterspeech
CSO	Civil Society Organisation
CT	Counterterrorism
CTR	Click Through Rate
EU	European Union
ISD	Institute for Strategic Dialogue
P/CVE	Preventing and Countering Violent Extremism
RAM	Reasoned Action Model of Behaviour Change (also known as TRA/TBP)
RAN	Radicalisation Awareness Network
SoRaSCo	Socialisation and Reasoned Action Strategic Communication Model
SR	Search Redirect

TPB	Theory of Planned Behaviour (modified name for TRA and RAM)
TRA	Theory of Reasoned Action
VEO	Violent Extremist Organisation

List of Figures

Fig. 2.1 Research areas relevant to this book 13
Fig. 2.2 The Reasoned Action Model (RAM) 36
Fig. 4.1 Adversary analysis (Strategic Framework) 98
Fig. 4.2 Effects framework (for counter narrative product—Example 2) 104
Fig. 4.3 Sub-effects and supporting effects 106
Fig. 4.4 The reasoned action model 107
Fig. 4.5 RAM—X1's contextual factors, attitudes and behavioural beliefs 109
Fig. 4.6 Normative beliefs and reasoned action model 110
Fig. 4.7 Control beliefs and reasoned action model 112
Fig. 4.8 Behavioural intention stage 113
Fig. 4.9 RAM and counter narrative interventions 115
Fig. 4.10 AX marketing funnel approach 119
Fig. 4.11 Targeting, sentiment and custom messaging 120
Fig. 4.12 Search engine results—raising awareness of the AX brand 121
Fig. 4.13 Action stage—raising awareness of the AX brand 122
Fig. 4.14 Contextual keyword term display ads 123
Fig. 4.15 AX YouTube Ad serving 124
Fig. 4.16 AX campaign results 127
Fig. 4.17 AX and SoRaSCo model analysis 130
Fig. 4.18 AX refutation 1 132
Fig. 4.19 AX refutation 2 132
Fig. 5.1 Reasoned action model 143
Fig. 5.2 CC and SoRaSCo model analysis 148
Fig. 5.3 Search Redirect and SoRaSCo model analysis 154

Chapter 1
Introduction

Abstract This chapter introduces the central focus and conceptual framing of the book's inquiry into the role, efficacy, and foundational assumptions of online strategic communications within the Preventing and Countering Violent Extremism (P/CVE) sector. It situates the discussion in the evolving digital environment, where counternarratives—messages and campaigns designed to derail pathways toward extremist violence—are disseminated to "vulnerable" and "at-risk" online audiences. The chapter posits that while these initiatives leverage sophisticated technological tools and data-driven approaches to influence audience behaviour, their underlying theoretical and conceptual bases often remain under-examined. Drawing on theories of radicalisation, deradicalisation, socialisation, and behaviour change, the chapter introduces a new evaluative framework, the "socialisation and reasoned action strategic communication" (SoRaSCo) model, to more rigorously test the strategic claims and asserted successes of P/CVE counternarratives. Ultimately, this chapter establishes the research parameters, theoretical underpinnings, and methodological approaches that guide the book's subsequent empirical and analytical work.

Keywords CVE—countering violent extremism · PVE—preventing violent extremism · Counter extremism · Communication · Strategic communication

The purpose of this book is to examine the use of communications efforts in the prevention and challenging of extremist narratives online. These efforts are based on the expansive and well-resourced sector of preventing and countering violent extremism (P/CVE herein). They leverage technology (tech herein) tools (algorithms, direct messaging, and audience targeting) to communicate counter-extremist messaging across various digital online platforms. This book focuses its research and discussion on the role of strategic communications efforts within this space. Specifically, the focus is on those efforts that are designed to deradicalise and change behavioural features of 'vulnerable' and 'at-risk' audiences who are deemed susceptible to extremist messaging. This assertion views the process of radicalisation-based pathways towards violence as a critical issue to tackle within online P/CVE communication efforts.

M. Zamir and C. Kaunert, *Strategic Communications' Role in Counter-Terrorism*, Advanced Sciences and Technologies for Security Applications,
https://doi.org/10.1007/978-3-031-89110-6_1

The role of communication in this context takes on a strategic guise. This is due to the assumption behind such messaging. P/CVE online efforts include objectives that seek to reduce, undermine, and divert at-risk audiences away from harm. This can be observed as creating a form of intent behind the messaging. As such, this sector and its work in the online sphere traditionally fall within the general paradigms of strategic communications campaigns found in the corporate world and the needs of health and social issues management. The central objectives in such contexts usually contain a desire or intent to create a change in audience behaviour. The P/CVE sector has steadily grown into a mass communications and prevention space. It also includes elements of military psychological operations (psy-ops herein), advertising industry tools of persuasion and influence, and a form of virtual community management. These elements have merged over the past decade into a branch of work called 'counternarrative' or 'counterspeech' communications.

Within the contemporary social structures of the digital age, the concept of human agency requires a refined understanding. Traditionally viewed as a binary state possessed by 'developed' societies and individuals, agency now presents as a complex interaction between individual volition, the dense network of social connections, and the pervasive influence of digital technologies. Once a relatively stable construct, identity becomes a dynamic negotiation shaped by online communities, algorithmic curation, and the constant performance of self in the digital sphere. Boundaries of belonging and loyalty transcend the physical world, transforming into a multifaceted interplay between physical and virtual spaces. Shared narratives, perceived intentions, and the intricate dance between online and offline interactions dictate the formation and reformation of allegiances. The lines between social norms and individual desires become increasingly blurred in this dynamic environment.

Traditional models of strategic communication, built upon the premise of a passive audience, need to be more robust. Human agency is not a target for manipulation; instead, it is the ever-shifting foundation upon which narratives of choice and volition are constructed. To navigate this new social landscape effectively, communication strategies must acknowledge the evolving nature of human agency and craft messages that resonate with the interconnected individuals who define the digital age. This premise is pivotal to the motivation behind this research project and its desire to unpack the relationship between communication, human agency and behaviour within online P/CVE efforts.

The book builds upon two decades of direct practitioner experience delivering P/CVE efforts online and offline. Often actioned through convoluted governmental policy around preventing violent extremism, terrorism and shared values, a key frustration has emerged around how strategic, organised and effective such efforts have been. It has become a terrain besieged with conceptual conjecture but an evident lack of evaluative and competitive credence. This is most evident in the efforts known as online counternarratives. Such work has mainly been carried out assuming that forms of communication delivered to mass audiences within which some members may be vulnerable to being radicalised can, in some way, shape or form, reduce this risk. In essence, this 'way' is known as a desire to alter behaviour back to a socially acceptable level or, put another way, reduce the likelihood of harm through communication. Such efforts merge with the idea and reality of human agency in the online space, enabling a detailed investigation of these assumptions.

This book seeks to understand and assess such efforts' role, efficacy, and implications within the broader counter-terrorism (CT herein) landscape. It uses UK Counter Terrorism efforts (via the Prevent Strategy) as its primary contextual source. This will be achieved by applying socialisation theory as a central analysis tool. This contrasts with what has become a psychology-centric paradigm of seeking behavioural change. This is often pursued without considering the social role of communication, narrative, normative alignment and human agency. Socialisation theory will be applied to the methodology by assessing the effectiveness of behaviour change-oriented communications using a specialised model designed by the researchers. This incorporates the theory into behaviour change understanding using the reasoned action model (RAM) and layering relevant elements of communications theory to supplement the fluidity of analyses where needed. The model is referred to in this book as 'socialisation and reasoned action strategic communication' (SoRaSCo).

Originality and Significance

The assertion under which this book is carried out is that a paradigm shift occurred in the early 21st Century. This move is responsible for the pedagogy and epistemology of terrorism studies so prevalent since 9/11, which have shifted from a more holistic understanding to a psychology-centred one. The subsequent birth of the P/CVE industry is a manifestation of this change. Has this paradigm shift been intentional? Has it ignored the role of human agency, faith, theology, and specifically, social processes such as socialisation in a pursuit to 'empiricise' the discipline and the entire discourse on how terrorism is understood? Through narrative and discourse analysis of case studies, the assumptions behind the reasoned action model (RAM) of behaviour change, P/CVE approaches to counter-extremism needs and the overarching reliance on data and technology can be critically examined. This research project bears the hallmarks of a 'health check-up' for over two decades of intense and formulaic P/CVE-based work across the globe.

This book seeks to break P/CVE online efforts down to their communicative parts. Using strategic communication as an essential element, has P/CVE been able to evidence its uses beyond serving its industry interests, or are P/CVE's strategic aims not as strategic as they initially appear? The emerging knowledge suggests that P/CVE is countering its presence by failing to be dynamic and nuanced in its approaches. If this is the case, then counternarratives are at risk of being counterintuitive. Against the backdrop of a rapidly changing virtual and social world, the evolution of how human agency is manifested and changes to existing understanding of citizenship, democracy and the rule of law are now everyday transient issues [4]. Ideas, values and notions of representation are now party to a trial by groupthink, disinformation, cancel cultures, hybrid social movements with violent tendencies, and a general decline in trust levels of traditional institutes of Statecraft and governance.

In adopting a new methodological approach to this book via the SoRaSCo model, it is hoped the findings shed new and more nuanced light on the P/CVE sectors' efforts to address the proliferation of extremist content online through counter-narrative campaigns. Analysis of the strategic nature and conceptual foundations of P/CVE efforts can inform future research and broaden the field's scope.

State of the Art

P/CVE was introduced to the UK security lexicon in the aftermath of the major terrorist attack on July 7th 2005, known as 7/7. Islamist extremism has taken both domestic and global forms, culminating in a coordinated attack on transport networks in central London (more details from news source). Loosely connected to the modus operandi (at the time) of Al-Qaeda and offshoot affiliates such as Al-Muhajiroun, this attack focused on already existing domestic counterterrorism efforts to address Islamist threats. In contrast, previously, this had been relevant to the Irish terrorist context [6, 7]. Much of P/CVE work has focused on programs targeting specific audiences with specific messages. This has taken the form of community-based interventions to safeguard vulnerable/radicalised individuals in the offline domain and online counter-extremism communication efforts. In a post-9/11, post 7/7 and post-Daesh Caliphate world, the overwhelming majority of counter-extremism and counterterrorism efforts have focussed on the Muslim community, targeted both as a risk community and a potentially violent threat [8].

CT efforts include various policy and strategy areas, including security infrastructure, intelligence gathering, communications efforts, community engagement and prevention. The rise of P/CVE communications-led approaches has created curiosity and conjecture concerning what these efforts seek to achieve and how and to which audience. Two specific areas of interest are the conceptual assumptions on which such counternarrative efforts are based and the framework under which these interventions are considered 'effective' or successful. Regarding ascertaining conceptual matters, radicalisation studies are particularly relevant, given its emphasis on pathways and processes that lead people towards violence. Equally, deradicalisation discourses also offer proximity to the work of counternarratives, given the underlying assumptions that communications efforts can positively affect already radicalised individuals by contributing to a process of diversion away from harm. This process and assertions are closely linked to conceptual understandings of deradicalisation studies. Radicalisation and deradicalisation discourses are central elements from which online communications efforts seem to take direction and inspiration. This book will examine the relative merit of this approach in line with the research aims. The issue of what constitutes an effective or successful counternarrative communications effort hinges mainly on how the stakeholders and creators of such content define outcomes. This leads to many possible criteria, from general social media 'vanity metrics' (likes, shares, reach, impressions) to complex behaviour change models being utilised. Both these elements will be studied in detail to ascertain how helpful behaviour change models and vanity metrics are in the battle against extremism in the online space. The research criteria will include a detailed look and assessment of the most well-known behaviour change approach, the reasoned action model (RAM herein), which merges the theory of reasoned actioned (TRA herein) and planned behaviour (TBP herein) into an overarching framework. In studying the conceptual and outcome-based elements of counternarrative communications efforts, it is hoped that this project will uncover how strategic such work is, what makes it practical and how it might be improved. Such scope considers the rise of the tech sector and its collective power to track, measure and engage diverse audiences. Leveraging tools such as sentiment analysis, online search redirect methods, direct messaging and user journey tracking are now standard operating protocols in this sector.

Theoretical Framework and Approach

This book aims to evaluate the body of work known as P/CVE counter-narratives through the lens of socialisation and strategic communications linked to behaviour change approaches. It also seeks to offer a framework for evaluation that is, in part, critical while highlighting best practices and possible pathways for future work. Because of the P/CVE sector's emphasis on online counter-extremism, aimed primarily at dissuading specific audiences from adopting or supporting extremist ideas, the main focus of this book will assess methods, frameworks and processes involved in efforts to undermine narratives of hate, extremism and violence, which have evolved into terms such as 'counterspeech', 'alternative narratives' and more broadly, 'counternarratives'.

The frameworks of socialisation, behaviour change (RAM) and communication theory frameworks will be operationalised through a detailed study of relevant literature and research. This will be achieved via a thorough and nuanced exploration of pertinent case studies from the P/CVE sector. These examples have sought to harness online tools and techniques to reduce audience vulnerability to extremism and violence. In doing so, the role and use of narrative, behaviour change models, measurement and evaluation elements will be considered to provide a developmental framework for this work moving forward.

This task was made easier by honing in on one significant element of empirical controversy within CT and P/CVE discourses. The continued reliance on radicalisation to predicate and build policy and practice within this sector offered an ideal vantage point for developing a robust analysis of the efficacy, effectiveness, and malleability of P/CVE and its communicative elements. From this foundational point, two distinct theoretical frameworks emerged that possessed the capacity to explore, identify and critically analyse the subject area. Crucially, these frameworks offered an element of specificity and access to broader contextual paradigms, which could be applied to radicalisation discourse and strategic communication concepts.

The two frameworks chosen were the reasoned actioned model (RAM) of behaviour change and the socialisation theory.

The RAM approach has been floating in and around P/CVE strategic communications and has yet to be fully contextualised, adapted or developed. As such, much of the efforts created by CVE strategic communications efforts, loosely known as 'counternarratives' are based on findings derived from specific social issues such as health management, sexual health communications and broader contexts such as commercial advertising practices. It appears this model is used within CVE counter-narrative design without a specialised or adapted version. This matter is exacerbated by much of this counternarrative effort being delivered in the online domain. Therefore, the context and sheer depth of factors related to social media/online communications and CVE become relevant issues for such a research enquiry. The theory of socialisation provides a broad depth of field to assess the validity of radicalisation discourse. As both a critical analyser of the assertions found within radicalisation studies and as a potential alternative to its claims, socialisation theory offers both possibilities for future development needs while also serving as a critical scoping tool in evaluating what this discourse is seeking to explain. The online context of this issue (online extremism) also means that using and championing data-driven, big-data, and sentiment analysis approaches has become part of the fabric of CVE strategic communications understanding.

Whilst appearing innovative and earnest in its auspices, the question that needs to be answered is: What foundations of understanding is the data-driven P/CVE approach built upon? The answer is that P/CVE has come to depend on what is known as 'vanity metrics' as a measure of 'success', behavioural change indicators as evidence of effectiveness and sentiment analysis as a conduit to understanding the nature of audience engagement. To adopt a more nuanced approach to understanding the assertions, process and practices of online P/CVE strategic communication, a method of investigation was needed that was capable of capturing the conceptual and operational considerations involved. The methodology for this book was chosen to use qualitative discourse and narrative analysis via case studies. In addition, the SoRaSCo model was utilised to analyse the data in depth. This would technically enable a contrast to quantitative approaches to P/CVE, which are often built upon unqualifiable assumptions about the nature of extremism, persuasion and behaviour change. Uncovering why such claims are made would be a key outcome of a study with this framework in place.

A critical factor in the quantitative approach to P/CVE study, measurement and evaluation needs is the emphasis on seeking a set of scientific causal elements with which to be able to monitor trends, assert predictive analysis on what drives extremism, and apply statistical behavioural change modelling to potential counter-extremism methods [5]. This 'empiricalisation' of terrorism studies, counter-terrorism efforts, and risk/threat analysis has coincided with a noticeable shift in governmental approaches to security policy. The emergence of a highly securitised mindset among European and broader Western nations has manifested most strikingly through the classification of security policy, built on a threat-based understanding of the likelihood and associated risk posed by violent extremism/terrorism-related trends in Western nations [9].

Charting a historical course from Martha Crenshaw's well-known 'Rational Actor' model of terrorism, the gradual move towards rationalising both the risk and the response to a 'rise' in threat perception from terrorism-related issues has spawned a CVE industry determined to prove its propositions in a manner akin to the famous theory of Karl Popper, namely falsification and verification [1–3]. Terrorists 'make choices', and these choices can be categorised, quantified, cross-referenced and then measured using predictive tools around the nature of persuasion, behaviour change and influence metrics [10]. To understand the penchant for favouring this approach, a methodology able to unpick the assumptions behind these processes was required.

In using discourse and narrative analysis, the claims made by counternarrative practitioners, P/CVE efforts and researchers in the field could be critically examined in two ways: firstly, by outlining the conceptual flaws in radicalisation theories and, by default, the majority of P/CVE counternarrative as a result. This would highlight the inherent weaknesses of the P/CVE approach to the casual processes associated with extremist recruitment and its vulnerability. Secondly, this methodology would allow a robust examination of the actual communications, products, and initiatives delivered by P/CVE efforts. In other words, it is important to assess the level of strategic intent and assertions of success and effectiveness made by such efforts. Discourse analysis is an ideal way to explore, understand and highlight positive and negative elements of online counternarrative strategic communications within P/CVE. This approach also offers a window to open the 'pandora's box' of how socialisation processes, social dynamics and social structures appear to have been shelved as useful conceptual tools in favour of the cognitive/empirical approach mentioned earlier.

The rationale of this research hinged, in essence, on identifying why these changes and approaches have transpired in this manner. The research proposition suggests the answer could lie in a subtle shift in how scientific enquiry is conducted, a dramatic change in how information is understood, or a combination of both. Effective research requires considering how existing messages are strategically designed to influence audiences. This means understanding the goal of the communication and its impact. As such, the research project investigates the development of novel online communication strategies to counter extremist messaging effectively. Such strategy must deliver accurate information and employ captivating elements to win over the target audience.

Summary of the Methodological Approach

Specific controls were needed to allow for the chosen methods to be applied in a manageable way. A qualitative discourse analysis via case studies was most suited to this book. It was decided to limit the research scope to analysing counternarratives aimed at challenging Jihadist ideology during and around the zenith of Daesh recruitment circa 2014–2017. This choice was also predicated on the researcher's own fieldwork experience and subsequent access to sources of information on online and offline P/CVE interventions that would facilitate a high level of detail and depth concerning being able to conduct critical discourse analysis. Another factor in this decision was the lack of publicly available information on CVE counternarratives beyond 'evaluation' reports from project teams or third-party think-tank organisations such as the RAND co-operation, Hedayah, Radicalization Awareness Network (RAN) and Institute of Strategic Dialogue (ISD). It was necessary to use information and knowledge from offline interventions that allowed detailed analysis but stayed within an anonymised and, therefore, sanitised data approach from a research and ethical standpoint.

The aim was to assess if P/CVE online counternarratives can claim they are practical and intuitively designed to tackle the problems they claim to. A vital element of this methodological approach is the ability housed within discourse analysis to bring into focus specific conceptual analysis and broaden debates with broader social dynamics and interpretive techniques. Also, discourse-based narrative analysis enables ideas, conceptual development, cultural factors, social interaction processes and comparative debates to occur within an overarching interpretive context. This suite of tools can foster implied and reactionary meanings to be understood, contributing to an overall understanding of how effective a specific narrative can or cannot be. This book offers industry-specific analysis by including conceptual and practical discussion of strategic considerations in communications design and shows how well this applies to P/CVE efforts.

The case studies chosen were the Abdullah-X (AX) campaign, Facebook's (Meta) Counter Conversations (CC) project and Google's Search Redirect (SR) method. These initiatives share common threads in the context of being ground-breaking counternarrative P/CVE efforts, using technology tools/platforms to amplify or enhance their presence and design elements that point to the use of critical strategic communications processes. As such, a detailed and nuanced analysis of the thinking, application, and lessons learned from these endeavours offers a roadmap for understanding the role of strategic communication in P/CVE efforts. These examples also allow for the cross-analysis of how predominant discourses within P/CVE, such as radicalisation process assertions, behaviour change modelling, and predictive risk-threat dynamics, have become so favoured.

Therefore, the key to this project's success as a research effort hinges on identifying the basis for this P/CVE approach and its relative effectiveness through scrutinising issues/claims of efficacy based on robust critical discourse analysis. A thorough examination of the tendency of this sector to use cognitive-behavioural (psychology-centred) approaches to understanding journeys into and out of extremism, the facets of how and why behaviour changes, and the role of influence and persuasion in online and offline intervention settings should it is hoped, allow for some new conceptual threads to emerge with such results produced as a result of this book, the field of online P/CVE interventions/counternarratives could in theory, further its 'craft' and conceptual depth with new learnings and perspectives.

Below is an outline of chapter structures to assist the reader in navigating the book. Chapter one begins with a review of extant literature.

The introduction chapter introduces the research aims and summarises the theoretical, methodological and contextual issues associated with P/CVE strategic communications counternarratives.

The second chapter focuses on the extant literature and knowledge from discourses on terrorism studies, socialisation theory, and behaviour change models. Its aim is to thoroughly document how radicalisation and deradicalisation discourses have become prevalent, the potential usage of socialisation theory as a conduit to understanding extremist pathways, and the significant contribution of communication-based behaviour change models by reviewing the RAM approach.

The third chapter enhances the literature review by operationalising socialisation and political socialisation theory and communicative behaviour change studies to create and introduce an overarching research framework for studying counternarratives: the 'socialisation and reasoned action strategic communications' model (SoRASCo).

The fourth chapter richly details how the SoRaSCo model is applied to the Abdullah-X animated counter-narrative project. It utilises all elements of the research model to understand and evaluate the project's facets, features, limitations, and outcomes.

The fifth chapter studies Meta's Counter Conversations project and Google's Search Redirect project to build out areas of commonality and difference as understood by the SoRaSCo model used in this book.

The sixth chapter offers broader findings and an assessment of efforts within the P/CVE context of online communications. It focuses mainly on the prevailing approaches of such work, which include design, audience targeting, delivery, measurement and evaluation needs. In doing so, it offers a robust assessment of the sector's efforts. It pushes the case study findings from the preceding chapters towards an overall contextual understanding of pertinent issues.

The concluding chapter brings the book's main findings together. It provides both summative and critical discussion of the state of the counter-narrative field, areas for improvement and future research development in the sector. A key argument presented is that counternarratives may prove counter-intuitive if their strategic intent, conceptual assertions and evaluation frameworks continue to overlook the role of socialisation and nuance among assumed 'homogenous' target audiences.

References

1. Best, S. (2003). *A beginner's guide to social theory*. Sage.
2. Crenshaw, M. (1981). The causes of terrorism. *Comparative Politics, 13*(4), 379–399.
3. Crenshaw, M. (1995). *Terrorism in Context*. State Press.
4. Fukuyama, F. (2019). *Identity: Contemporary identity politics and the struggle for recognition*. Profile Books.

5. GIFCT. (2021). *Content-sharing algorithms, processes, and positive interventions working group Part 2: Positive interventions.* [online] Available at: https://gifct.org/wp-content/uploads/2021/07/GIFCT-CAPI2-2021.pdf [Accessed 18 Dec. 2022].
6. Hickman, M., Thomas, L., Silvestri, S. and Nickels, H. (2011). *'Suspect Communities?' Counter-terrorism policy, the press, and the impact on Irish and Muslim communities in Britain.* [online] openaccess.city.ac.uk. Available at: https://openaccess.city.ac.uk/id/eprint/8735/
7. Hillyard, P. (1993). *Suspect community.* Pluto Press (UK).
8. Krueger, A. B. (2019). *What makes a terrorist: Economics and the roots of terrorism.* Princeton Univ Press.
9. Rhinard, M. (2019). The crisisification of policy-making in the European Union. *JCMS: Journal of Common Market Studies*, 57(3). https://doi.org/10.1111/jcms.12838.
10. Saltman, E. and GIFCT. (2022). *Introducing 2022 GIFCT Working Group Outputs.* [online] Available at: https://gifct.org/wp-content/uploads/2022/07/GIFCT-22WG-ExecSummary-1.1.pdf [Accessed 21 Mar. 2023].

Chapter 2
Radicalisation and Counter-Terrorism—Reviewing the Problems in Strategic Communications Role in Counter-Terrorism

Abstract This chapter critically re-examines the prevailing discourses and conceptual frameworks underpinning radicalisation, deradicalisation, and their relationship to counter-terrorism efforts, with a particular focus on the strategic communications dimension of Preventing/Countering Violent Extremism (P/CVE). It argues that current understandings of radicalisation too often rely on psychology-centric models and ambiguous, under-defined constructs that struggle to account for the complex, socially embedded processes by which extremist ideas and identities take shape. Drawing on classical sociological theories of socialisation and human agency—particularly as articulated by scholars such as Parsons, Giddens, Bourdieu, and Bauman—this chapter highlights the importance of understanding extremism as not just an individual cognitive shift, but also as a re-socialisation into alternative normative frameworks. In doing so, it makes a case for moving beyond linear and universalist radicalisation models towards more nuanced, context-sensitive approaches that acknowledge how digital ecosystems, fluid cultural norms, and rapid technological change have transformed both extremist recruitment and its prevention. Additionally, the chapter explores the limitations of applying behaviour-change paradigms, such as the reasoned action model (RAM), without accounting for the complexity of collective identities, community-level narratives, and socially constructed meanings.

Keywords Counter extremism · Communication · Extremism · Reasoned action model · Cognitive behaviour · Strategic communication

This book uses socialisation theory, behaviour change models (RAM), and (strategic) communications approaches to analyse the effectiveness of the P/CVE online communications sector. Adopting this approach, the research study and its parameters take on existing sector knowledge with a unique and original perspective. This is because previous research into P/CVE communications has been based either on behaviour change evaluations or psychology-centric ontology, as well as hinging success criteria on mainstream social media analytics (vanity metrics). What makes this book original and significant in this regard is its use of classical sociological theory (socialisation) merged with behaviour change models and then layered

M. Zamir and C. Kaunert, *Strategic Communications' Role in Counter-Terrorism*,
Advanced Sciences and Technologies for Security Applications,
https://doi.org/10.1007/978-3-031-89110-6_2

through the lens of strategic communication. It seeks to understand both the social aspect of communication and behavioural aspects of what counternarratives within P/CVE can and cannot feasibly hope to achieve.

As such, four primary bodies of knowledge should be understood to comprehend the broader comprehensive research and analysis approach of this book: (1) socialisation/political socialisation theory, (2) radicalisation and deradicalisation theories, (3) strategic communications influences on behavioural change, and (4) existing literature on the P/CVE sector. Whilst strategic communications in the marketing and advertising sector are a field of work and development in their own right, less of its learning and conceptual underpinning applies directly to social science discourses within P/CVE. This results in needing to broaden the field to have a P/CVE-specific focus. The 'social' element of extremism, narrative, and cultural reproduction points to a need for communications and P/CVE to be understood more in terms of dynamic and exchange-based interactions within social/virtual settings. One could suggest a sociology of P/CVE as a better term, and socialisation theory seemingly offers an ideal segue into this context.

Political socialisation theory is the umbrella theoretical approach to understanding how and why individuals choose to align with specific sociopolitical identities. Therefore, natural sub-sections of political socialisation are radicalisation and deradicalisation theories, analysing how and why individuals (or groups) might align with hate-based, violent extremist ideologies or turn away from such ideologies. From these theoretical lenses, we see the practical application of P/CVE strategies through strategic communication for proactively trying to inoculate parts of society from violent extremist ideologies or guide audiences away from violent extremist groups. In the last fifteen years (approx.), there has been an increasing body of literature measuring and evaluating P/CVE approaches. Lastly, and linked to the previous point, the ultimate goal within P/CVE approaches is to cause a, hopefully measurable, sentiment or behavioural shift in an audience away from a hate-based and violent ideology. A base understanding of behavioural change theory is necessary to guide measurement and evaluation approaches and conclusions about the booming or lacking success of a given P/CVE deployment. The objective of this chapter is to offer an in-depth understanding of the approaches mentioned above relevant. It is structured along the logic flow outlined in the diagram below (Fig. 2.1).

Socialisation Theory

A relevant and critical place to begin this review is the theory of socialisation, which explains how individuals acquire their 'role' by fulfilling the requirements of such roles as defined by the prevailing social system. By and large, this approach has a normative component, which loosely translates as the belief that once socialisation is 'attained', subsequent manifestations will always be evaluated in light of its overall beneficial effect [98]. This understanding is central to the functionalist perspective in sociology, which views socialisation primarily as a means of teaching

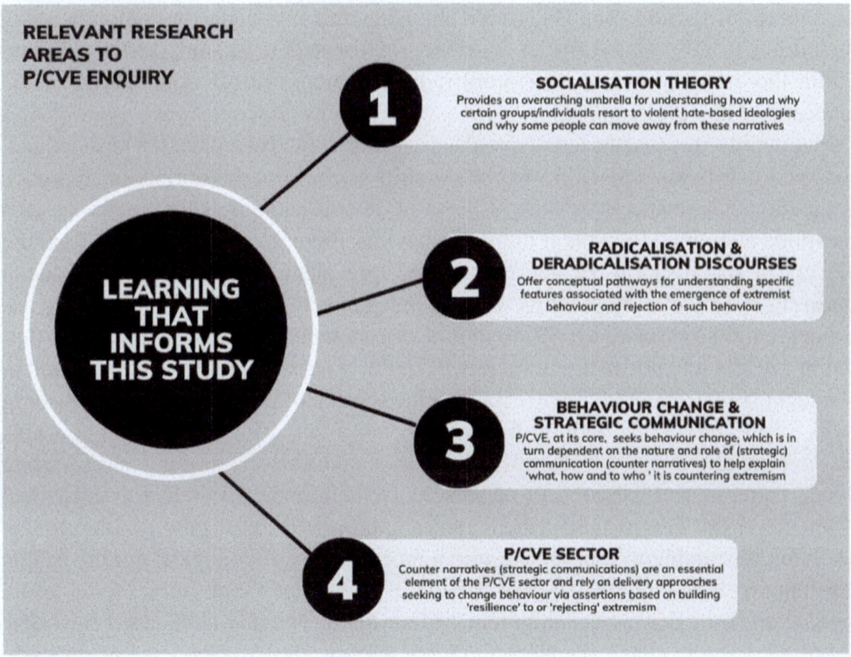

Fig. 2.1 Research areas relevant to this book

people how to live in their social worlds as part of an overarching system of dependent and 'functional' systems collectively referred to as 'society'. The common thread here is that socialisation educates people about their 'place' in an industrialised landscape typified by the advent of modernity [19, 27, 72, 74]. In contrast, the term 'terrorism studies' applies to at least three distinct disciplines. This wide berth in usage benefits this particular research endeavour, given that much of the prevailing discourse on terrorism studies originates in social psychology (with a stronger emphasis on psychology), criminology, and security studies [41]. What terrorism studies discourse, particularly the CN field, clearly lacks are approaches that seek to understand the distinctions between social process perspectives on the issue of radicalisation. Instead, the focus appears more on behavioural action-oriented concepts that underpin strategic communications and counter-radicalisation work.

From the outset of this research enquiry, radicalisation appears to be a term that attempts to supplant the concept of socialisation. It does this by claiming to be more conceptually appropriate for the emergence or mechanism by which an individual's mindset can adopt a greater proclivity for violence. Choosing radicalisation instead of socialisation in this fashion provides points of research-based entry into both social processes and cognitive behaviours [54]. Thus, to delve deeply into the field and role of strategic communications within P/CVE, it is necessary to identify gaps in and underlying assumptions about what constitutes 'effective' efforts versus those deemed ineffective, as well as the justifications for these claims. If, on the one hand, the concept of socialisation is a viable and adaptable paradigm for comprehending the journey from 'mainstream' to 'extreme' mindsets, then its study and application within this review of extant literature require a more critical examination of the concept and its development.

One could argue that the term 'socialisation' was coined to replace the term 'education' and that it was created to express a shift away from the normative approach of functionalist perspectives such as those of Talcott Parsons. Thus, sociology as a discipline attempted to explain human behaviour more empirically by the time the concept of socialisation became increasingly prevalent in discourses about social worlds [27]. However, both terms were intended to express and comprehend the use of intervention by 'social agents' to mould and create individuals capable of participating actively and positively in society [44]. Early research in this field appeared to be caught between attempting to establish the science of personality by establishing a close epistemological relationship with socialisation, even though both concepts are conceptually distinct. The primary goal of socialisation is forming and adopting shared patterns, and the concept of personality is intrinsically linked to individual differences [45]. This explains why Parsons' approach to socialisation was critical in resolving this ambiguous approach into a more concise sociological narrative. It is worth noting here that the idealised version of Parsonian socialisation served as the foundation for a slew of developmental ideas that eventually challenged this view and offered a more nuanced understanding of the concept as society entered the early twentieth century, characterised by rapid industrialisation, war, and suffering [13].

Exemplified by Merton, Mead, Becker, Goffman, and Garfinkel's works, sociology and the study of socialisation developed new conceptual frameworks. This resulted in a shift away from empiricism toward a more interpretive understanding of social phenomena and processes. Merton's analysis of Parsons' 'positive socialisation' introduces the concept of 'anomie', which implies that any prevailing social success goals become dysfunctional for members of society who lack institutional means to pursue them [87]. Both Danziger, Brim and then Wheeler argue that immigrant societies suffer from this characteristic in terms of a general tendency among disparate groups to be unable to socialise into the host society's dominant norms and values for structural and situational reasons [19, 27]. Such framing hints at the role of the psychological or cognitive effects of such processes on people, which in turn gave rise to social psychology as a discipline in its own right. This is an appropriate time to clarify the fundamental premise of what classical theories of socialisation suggest is the processes' ultimate goal.

Socialisation does not exist unless there are identifiable and attached sets of norms and values into which various formal and informal agents of socialisation process individuals. From the family, school, peer group, and mass media, socialisation refers to the process by which roles are learned concerning social structures and expectations [116]. However, the apparent disadvantage of this classical approach is that the process is transmissional in nature (from socialising agent to individual) rather than transactional. The dominant society's norms and values are offered to the populace in the hope that they will be accepted and then adopted and reproduced through amenable social structures [9]. The initial concept of dysfunction is introduced through the concept of 'anomie', and as the field developed, other variants of a more transactional narrative of socialisation emerged.

Developing new theories about the individual's role and place in the social order was based on interpretative understandings of how identity, meaning, and social life are negotiated and managed. Perhaps it is with these social action-oriented lenses applied to radicalisation and socialisation discourses that one of the central arguments of this research study forms: socialisation into mainstream society, dysfunction from these mainstream values, the offer of a 'new socialisation' via 'extremism', and the subsequent attempt to quell this new process are assumed to be at the heart of strategic communications. Currently, this 'logical framework' does not account for the security-centric lens of CT regarding how risk and threat narratives dictate policy and security matters through communication or the relationship between the nation-state, its foreign policy objectives, and its citizens. Equally, the discussion is currently only at a conceptual level. The enormous shadow cast by technological advancement, the age of social media, and the revolution in information access and sharing have not been adequately framed. This appears to be a necessary framework within which this entire field of work is taking place [5, 33].

To summarise, these factors significantly influence the overall outcome of contemporary socialisation efforts and, more specifically, the result of CNs. What, then, can these social action perspectives on socialisation theory reveal about the relationship between the individual, societal values, and extremism on a fundamental level? These questions become more pertinent in the digital age of social media influence and online extremism. However, given that P/CVE efforts are aligned with government national security policy, home affairs policy, and foreign policy, ignoring socialisation perspectives around social structures and their roles in individual and community dynamics would create a critical gap. This point brings a sharper focus to the breadth of literature and conceptual discussion needed to resolve the central question of the role of strategic communications in P/CVE CNs efforts and if socialisation theory, communication, and behaviour change studies are suitable approaches to address this need.

Thus, this book aims to examine the persuasive nature of P/CVE CNs. It will argue that the concept of socialisation and its variants (social action, human agency and structuration) have been overlooked in favour of behaviour change models that downplay the idea of the 'social agent', as advanced initially by Anthony Giddens in his structuration-based sociological approach, and further developed through the work of Pierre Bourdieu. A more holistic paradigm of factors can be presented in utilising socialisation theory as a foundational element in understanding why some people join extremist groups. This broader approach allows for the contemporary facets of social media influence and online identity development to be understood in a more theoretical context. The work of critical theorists such as Zygmunt Bauman also offers a powerful segue into the contemporary challenges of social agency through links to the fast-paced changes to social and technological processes that are now considered 'normative' everyday reality.

This perspective also lends itself well to the post-modern discourses, advanced originally by Jean-Francois Lyotard and Michel Foucault. The idea of a post-truth world and a dismantling of traditional socio-cultural and political structures and norms offers intriguing arguments about the role of power, authority and agency. Similarly, the seminal works of Francis Fukuyama on social change and political identity offer support to the idea that social change is not only a feature of a post-modern world but an active agent in the emergence of a more fragmented and lucid approach to identity, belonging and notions of loyalty and duty. This chapter begins, therefore, with a summary of the sociological approaches to socialisation and social agency. This will enable and lead to a more comprehensive look at how radicalisation and deradicalisation discourses have developed alongside the role of strategic communications within the P/CVE sector.

Anthony Giddens: Structuration Approach

Giddens posits that structuration pertains to the dynamic process by which social structures are generated, sustained, and altered due to the acts undertaken by individuals and collectives within a given society. According to Giddens, social structures are not merely external limitations placed upon individuals, nor do individual actions solely determine them. On the contrary, the author posits that there exists a symbiotic relationship between social institutions and human agency, wherein they interact and influence each other in an ongoing and reciprocal manner. The idea of structuration asserts that pre-established social structures affect people's behaviours, but simultaneously, their behaviours play a role in perpetuating and altering those structures. Social structures encompass a set of regulations, societal expectations, and available assets that influence human conduct. Conversely, human agency refers to the capacity of humans to actively and imaginatively understand and adjust to these social structures employing their actions. The crux of structuration theory is its emphasis on the dynamic interplay between structure and agency. It posits that individuals are not just recipients of societal influences but active participants who actively create and alter social reality via everyday practices and interactions.

The structuration theory and social agency approach developed by Giddens are intricately connected to socialisation since they provide a dynamic framework for understanding the reciprocal relationship between individuals and society in constructing social reality. The idea of socialisation places its emphasis on the mechanism by which individuals acquire and internalise cultural norms, values, and behaviours, thereby attaining the necessary skills and knowledge to participate as proficient members within a given community effectively. Giddens' theoretical framework supports this perspective by emphasising the reciprocal relationship between social institutions and individual agency. His framework recognises that socialisation is not a unidirectional process characterised by the passive acquisition of knowledge but rather a continuous interplay between individuals and the societal frameworks in which they are situated. Individuals engage in social practices and reflexivity to perpetuate and alter societal norms, thereby influencing their actions' underlying structures. This can be referred to as human agency. The dynamic relationship between individual human agency and the broader social context underscores the intricate nature of socialisation. It provides a more nuanced comprehension of how individuals contribute to society's ongoing formation and transformation (positive and negative).

Pierre Bourdieu: Social Agency and 'Habitus'

Bourdieu is another scholar whose work is connected to Anthony Giddens' concepts of structuration and social agency. Bourdieu's theoretical framework about social practice and habitus is a valuable addition to Giddens' conceptualisations, as it delves into the significance of socialisation and the intricate dynamics between structure and agency in influencing individual conduct. According to Bourdieu, people's behaviours are influenced by their 'habitus', a framework of dispositions and learned behaviours that are moulded by their social context and personal experiences [17]. These tendencies, in turn, are shaped by the broader social systems within which they are embedded. The theoretical framework developed by Bourdieu, which places significant importance on the internalisation of social structures and their manifestation through various practises, is a by product of Giddens' conceptualisation of the dynamic interplay between individuals and society. Both theorists emphasise the importance of agency in the reproduction and transformation of social structures, offering unique perspectives on the intricacies of social existence and how individuals traverse and engage with their social environments.

Critical Theory: Zygmunt Bauman and Modernity as a Key Social 'Agent'

The scholarly contributions made by Zygmunt Bauman link to 'critical theory', which emerged from the Frankfurt School in the mid to latter part of the twentieth century. Bauman's work is distinguished by his critical examination of prevailing power systems and emphasis on societal change. Bauman made substantial contributions to critical theory by examining modernity, consumer culture, and the notion of liquid modernity, establishing himself as a notable figure in sociology. Within the realm of socialisation, critical theory and Bauman's scholarship present a counterargument to conventional perspectives that posit a steady and amicable process of socialisation. Bauman emphasises the impact of dominant ideologies, media, and consumer culture on socialisation, illustrating how these forces can affect and shape individuals' beliefs and behaviours. Consequently, this process perpetuates repressive institutions and reinforces unequal power dynamics [9].

Post-Modernism and Socialisation

The post-modern conceptualisation of socialisation provides further intricacy to the concepts of structure, agency, and social process more broadly. Post-modernists critically examine conventional viewpoints by interrogating the durability and consistency of societal frameworks while refuting the notion of a singular, immutable identity. The authors contend that socialisation processes exhibit high fragmentation, diversity, and contingency as ever-evolving cultural, historical, and technical circumstances influence them. From a post-modern perspective, identities are characterised by their fluidity and diversity since various discourses shape them and their practices. There is a transition in focus from a unique socialisation process to a multitude of experiences and subjectivities. This perspective critiques the ideas proposed by Giddens and Bourdieu by dismantling the dichotomies between structure and action, person and society, and placing significant emphasis on the influence of power, language, and representation in moulding the fabric of social reality. Post-modern perspectives on socialisation advocate for a critical analysis of prevailing discourses and the acknowledgement of various marginalised voices, thus fostering a more comprehensive and inclusive comprehension of the intricate nature of present-day social existence. Critical and post-modern theories underscore the significance of reflexivity and awareness in effectively navigating the complicated dynamics of contemporary society.

Within the post-modern socialisation framework, a notable emphasis exists on fractured identities, the presence of numerous subjectivities, and the fluid nature of social institutions. This alignment with critical theory is rooted in its investigation of the mechanisms through which power operates and influences the lives of individuals. Bauman's theoretical framework of 'liquid modernity', which highlights the ephemeral and unpredictable characteristics of modern existence, aligns with the post-modern perspective that emphasises the ever-shifting and dependent nature of the social environment [9]. Nevertheless, it is crucial to acknowledge that critical theory and Bauman's body of work adopt a more overtly normative and transformational position than the post-modern approach. They promote the need for societal transformation, questioning and contesting repressive systems while striving for a society characterised by fairness and equality.

In contrast, the post-modern perspective on socialisation tends to use a descriptive and deconstructive approach, aiming to reveal the intricate and diverse nature of social encounters without imposing a predetermined image of what lies ahead. This emphasis links to the emergence of more lucid renderings of online identity and behaviours in a more 'tech-centric' society. The idea that narratives have replaced facts and taken the 'alternative facts' moniker evidences a more disparate and fractured approach to notions of 'truth' or universal facts [85]. Part of the rise of radicalised narratives and movements could thus be linked to the emergence of systemic changes in how modernity and society have clashed in different and nuanced ways. These issues, given their reliance on action, must also influence and alter notions of human agency as a by-product of such rapid social and digital change.

Prominent supporters of such a perspective include Jean-François Lyotard and Michel Foucault. Jean-François Lyotard is widely recognised for his seminal work titled 'The Post-Modern Condition', in which he delves into the fundamental aspects of knowledge and its intricate relationship with transitioning from modernity to post-modernity. The author presents the notion of 'incredulity towards meta-narratives', proposing that post-modernism challenges the comprehensive narratives or over-arching explanations prevalent in modern thinking (alternative facts vs universal facts). Conversely, Michel Foucault is widely recognised for his seminal contributions to power dynamics, knowledge production, and discipline systems. His scholarly pursuits in history, psychology, and criminology facilitated the formulation of the theoretical construct known as 'power/knowledge'. This construct highlights the intricate interplay between power dynamics and knowledge acquisition, underscoring their role in controlling and establishing regulatory mechanisms over individuals and society. This can be applied to the tension between mainstream societal norms and more radicalised norms created from the rejection of the former. The scholarly contributions of both Lyotard and Foucault have significantly influenced the post-modern theory field. Their enduring legacy is evident in ongoing dialogues concerning culture, society, and the intricate dynamics of modern existence [13]. Drawing on critical theory and Bauman's work, postmodern socialisation theory offers a nuanced analysis of contemporary social existence. It highlights socialisation's dynamic and multifaceted nature and the centrality of power and agency in social change. This framework helps to understand how individuals navigate identity and social circumstances in contemporary cultures.

Social Change and Political Identity—Francis Fukuyama

In the realm of contemporary social theory, Francis Fukuyama has gained recognition for his contributions to the fields of political science and international relations. Fukuyama garnered much scholarly and public interest by publishing his seminal work, 'The End of History and the Last Man', in 1992 [13]. Within the context of this literary work, the author posits the contention that liberal democracy serves as the culmination of humanity's ideological progression, signifying the ultimate conclusion of ideological conflict and, therefore, representing the 'end of history'. Fukuyama argues that liberal democracy, characterised by its prioritisation of individual rights and market-driven economies, is the most comprehensive system of governance that effectively fulfils human wants and aspirations [38].

Fukuyama's scholarly contribution examines philosophical and historical notions, primarily emphasising the history of political ideologies. The concepts put forth by Fukuyama exhibit a clear divergence from the perspectives of post-modernist scholars such as Jean-François Lyotard and Michel Foucault. The latter scholars prioritise deconstructing meta-narratives, interrogating assumptions surrounding stable identities and realities and examining the intricate dynamics inside modern society. In contrast, Fukuyama's scholarly contribution focuses on advancing political ideas and their potential ramifications for the future of human culture. In doing so, he provides a lens into possible causes for the rise in extremist ideologies, the breakdown in the relationship between the nation-state and citizens, and the re-emergence of populist ideas as 'reactionary' effects to contemporary social issues. These include migration, equality and diversity narratives and economic disparity within geo-political contexts. Extremist groups also use these factors (or variants of them) to amplify their respective causes. As such, Fukayama provides a compelling insight into how these issues have emerged in some form of politicised maelstrom of competing narratives around complex social issues [40]. Socialisation theory offers mechanisms to conceptualise and better understand this interplay.

Socialisation as a Segue into Radicalisation Discourse

Socialisation encompasses various theoretical viewpoints that collectively contribute to a thorough comprehension of human behaviour and societal processes. The theoretical framework developed by Anthony Giddens, known as structuration and social agency, elucidates the dynamic relationship between social structures and individual agency. This approach highlights individuals' active role in shaping and perpetuating social reality. This viewpoint follows post-modernist ideas that examine fragmented identities and the malleability of social institutions in the era of social media. Post-modernists critically question overarching narratives and emphasise the intricate nature of present-day society. Moreover, examining critical theory and the scholarly contributions of Zygmunt Bauman provide valuable insights into the impact exerted by power dynamics, prevailing ideologies, and consumer culture on socialisation processes. These observations lay the foundation for a stimulating investigation into the discourses surrounding radicalisation and deradicalisation. This discourse acknowledges the intricate nature of these occurrences, encompassing power dynamics, reflexivity, and individual action. It is crucial to comprehend the influence of social media and the period of individualism on the process of radicalisation and examine measures for deradicalisation.

This viewpoint corresponds to a broader issue concerning radicalisation that Matthew Francis highlights by reintegrating radicalisation into the socialisation narrative through social sciences, specifically sociology [37]. Additionally, Frank Furedi suggests that the move away from sociology in favour of psychological discourses, in his view, has set up P/CVE and CT discourses to fail due to a simple lack of real-world connectivity with the 'social'. In other words, radicalisation can never be a self-contained concept, which it could be argued has emerged as a standalone discourse post-9/11 [42]. Individuals acquire norms and values through interactions with various elements of society and its institutions and exercise human agency in critical life moments. Of course, self-socialisation is possible in today's digital world, but this still includes those who self-socialise into more radical ideas. Socialisation theory facilitates both an individual and communal element to norm acquisition, thereby housing traditional ideas around how people develop norms and more nuanced individualised versions. As a result of being asked to explain why people adopt extremist ideas and commit acts of violence, radicalisation has complicated a relatively simple social process and transformed itself into a 'hydra' figure of vilification both within and outside P/CVE circles [18, 54].

It is not implausible or illogical to view socialisation as a critical component of a more grounded understanding of radicalisation. The key to socialisation theory rests on its relative flexibility in approaching mainstream and more deviant narratives from a learned behaviour and interpretive cultural context. The theory offers a way to understand how people acquire norms and values and interpret, re-render and accept or reject them. It is a more comprehensive theory for understanding the levers and dynamics that ultimately shape and create more extreme or radicalised variants of social and political norms. It could be argued, therefore, that 'socialisation into radicalisation' is the logical pathway that appears to have been overlooked in P/CVE understanding since 9/11. The following section will introduce an in-depth review of the state of the field for radicalisation and deradicalisation discourses as relevant paradigms to the research aims of this study.

Radicalisation and Deradicalisation Discourses

This section outlines the fields of radicalisation and deradicalisation studies. It examines, highlights, and fleshes out extant concepts, assumptions, and mechanisms underpinning P/CVE and CT efforts in the post-9/11 context. Extensive academic efforts over the last two decades to capture the dynamics, nuances, and empirical reliability of this field means that the concepts of radicalisation and, by extension, deradicalization, remain highly contested and overly reliant on psychological renderings of the problem set [18, 54, 63]. As an element within this field, P/CVE CNs make extensive use of assertions derived from radicalisation studies and aspects of commercial advertising practice to attempt to create communications campaigns aimed at engaging target audiences in normative ideas that would ostensibly undermine extremist ideology and prevent or counter-radicalisation [48, 64].

A cursory examination of the policy frameworks guiding CT efforts reveals a need for greater distinction between security-oriented operations, integration-oriented policies, and the role of foreign policy agendas [52]. This hints at a clearer understanding of what CT even means. As a result, it is unsurprising that the EU's efforts in this area are frequently undermined by a lack of consensus among member states and difficulty in accurately framing the most critical problem sets [69]. This is exacerbated by a shortage of conclusive evidence regarding the specific goals of counter-radicalisation efforts (including communications), i.e., the 'end-state' objective of such efforts. Due to a distinct lack of understanding concerning the true meaning of deradicalisation versus disengagement [62], 'countering radicalisation' as a communications objective is invariably fraught with conceptual gaps.

Radicalisation: A Confusing or Contested Concept?

Examining the ambiguous meanings of 'radicalisation' and 'extremism' highlights the need to explore the origins of these terms and their role in shaping narratives surrounding moderation and extremism [10, 54]. This inquiry is crucial for understanding strategic communication in P/CVE efforts.

Disagreement about the definition of radicalisation hinders P/CVE efforts. Terrorism and extremism are subjective terms, making it difficult to reach a consensus on how to address them. This lack of agreement leads to controversy in the field. This is perhaps best exemplified by the continuing 'split' in perception between the 'environment' and 'ideology' camps concerning how best to frame the roots of extremism and radical beliefs. The most profound of these debates is arguably the public intellectual rivalry between Olivier Roy and Gilles Kepel. This vehement difference of opinion is based on the primary factors that facilitate the onset of radicalisation and violence. Kepel regards environmental and structural factors as primary causes, including Islam, for example, whereas Roy sees individual and behavioural (social-psychological) drivers as the means by which radicalisation emerges [54].

The respective discourses appear to be vying for the same arena space, oblivious to the fact that no individual affectation of 'self' (seeking) can exist in the public domain without the individual's construction of narratives and structures for interpretative rendering (narratives, ideas, and meanings); in other words, a combination of Roy and Kepel's theories. Through the interaction, exchange, and interpretation of this act, concepts of agency, choice, individuality, and community are then rendered according to desires, needs, and particular cultural-political systems [8, 11, 39, 45].

There is no environment without inhabitants, and no individuals can exist in the absence of a corresponding environment resembling a form of social structure [114]. This inability to find relatively common ground has a detrimental effect on how policy, practice, and evaluation are understood and implemented, as it is a well-established fact that once an intervention enters the theatre of reality, it becomes a part of the environmental conditions, producing first and second-order effects. The first-order effect is how VEOs react or alter their modus operandi, while the second-order effect is how specific communities support or oppose P/CVE efforts. The Kepel versus Roy disagreement points to why socialisation perspectives such as structuration, social agency, and social action are essential to broadening approaches within P/CVE that employ communication tools to prevent radicalisation and violence. This point can help facilitate a more thorough review of the extant literature on radicalisation studies compared to socialisation theory.

Lorne Dawson provides an excellent overview of the plethora of issues plaguing the radicalisation narrative in its entirety, from the gradual acceptance of how complex a discourse radicalisation is to the tensions between theoretical and methodological perspectives vying for space in the P/CVE and CT lexicons [29]. According to Dawson, one area that appears to be gaining consensus is the issue of a core set of parameters that still needs to be further understood, conceptualised, and tested. These factors include the following five issues:

- The Problem of Specificity
- The Transformation from Profiles to Processes
- The Importance of a Multi-Factor Approach
- The Problem of Heterogeneity
- The Issue of Primary Data

Apart from these five considerations, the fundamental issue with radicalisation discourse is that much of the debate focuses on the relative merits of various theoretical approaches on the one hand and critiques of methodological approaches on the other [29]. These efforts are not in vain, but they do not address the central criticism of radicalisation discourse, which is that, as a concept, it does not adequately address the needs of CT efforts. As Paul Hedges points out, "Perhaps the first step toward examining and comprehending radicalisation is to recognise that radicalisation is not an end in itself… I would argue that radicalisation is not a distinct practice or mode of thought" [54]. It is critical to understand each of Dawson's five emerging considerations to begin unpicking where the radicalisation discourse will go from here. Given the premise that it is attempting to explain socialisation processes, this requires critical analysis from a multi-level set of lenses to gain legitimacy within academia and operational understandings. The specificity problem seeks to narrow the range of possible causal factors that contribute to why some people turn to extremism and violence while others do not. Many of these factors do not appear sufficiently specific to qualify as 'core' factors. This issue continues to serve as the ultimate litmus test for the concept's efficacy and applicability in P/CVE efforts at all levels. P/CVE understanding has continuously shifted from profiles to processes, often aiming for generic radicalisation models. However, political narratives and profiling hinder this approach, potentially benefiting terrorist/violent extremist organisations (VEOs).

A multifaceted approach to P/CVE considers various factors, influences, and interactions contributing to violent extremism [61, 65, 113]. This approach avoids a one-size-fits-all solution. Both structural and motivating factors are equally crucial in comprehending the journey. This approach is gaining traction but has yet to become the prevalent mode of discourse. Similar to the issue of specificity, the issue of heterogeneity continues to be a concern. VEOs' methods of approach, dynamics, and networking systems are unique. This is true with regard to demographics, geography, political and cultural contexts, even before ideological differences are considered [18].

Finally, there is the issue of data, but concerns about the validity and reliability of the data sets still need to be addressed. There is still a pervasive tendency to rely on publicly available sources or the private testimonies of fighters, former extremists, or victims, both of which muddy the validity and reliability of trials [80]. Something that the discourse as a whole may need to consider is looking into those groups that do become extreme or adopt violence and determining whether specific dynamics are at work that may help explain why they, and not others, choose the extreme path. This would address the issue of specificity and limit the pervasive presence of profiling in CT practice and, to a lesser extent, P/CVE [53].

Despite the efforts outlined above to incorporate identified concerns into the developmental component of radicalisation discourse, critical questions remain about the 'value' of radicalisation as a concept. This is particularly relevant when closely linked to deradicalisation concepts and efforts. The central argument here is that if radicalisation as a concept is not yet 'fit for purpose', deradicalisation is also not. At the heart of this debate is how the threat posed by Jihadi ideology is currently understood through the application of radicalisation and deradicalisation concepts, which, when examined critically, appear either too broad to be effective or too narrow to accurately distinguish between religiosity and extremism on the one hand and terrorism and security on the other, when the concept and reality of strategic communications are applied to them.

Lack of Conceptual Clarity and Disparity of Radicalisation and P/CVE Contexts

Notwithstanding the abundance of academic work on radicalisation, much of this work overlooks the importance of radicalisation and deradicalisation as 'mutually symbiotic' phenomena. To be conceptually adequate, radicalisation discourse must address the entire process by which someone becomes attracted to extremism, what happens when they do, and how this can and cannot manifest. It must also be 'purpose-fit' when mapped against the concept of deradicalisation, as the existence of one is inextricably linked to the possibility of the other manifesting. When considering the relationship between P/CVE and strategic communications, much hope and expectation are placed on the notion that people can be persuaded, dissuaded, and influenced to avoid becoming radicalised or to leave this path via deradicalisation. This is arguably an oversimplified claim unsupported by empirical or narrative evidence over time. Additionally, it obscures both the human processes of critical reflection and first-level influence from close networks and issues of voluntary disengagement, desistance, and recidivism [63, 93].

There is an argument that radicalisation discourse lacks conceptual clarity regarding understanding how radicalisation and deradicalisation manifest within diaspora communities subject to CT legislation and ongoing media attention [99]. This precisely sits at the heart of this debate on the aims and objectives of P/CVE strategy and communications efforts. The primary purpose of this literature review is to examine the concepts of radicalisation and deradicalisation in contrast to socialisation theory. This aims to arrive at a point of understanding that enables detailed analysis of how P/CVE is strategically communicated, how it becomes integrated into the operating environments of both government efforts and adversary reactions, as well as the intermediary of the communities where the theatre of P/CVE efforts 'play out'. The concepts of radicalisation and deradicalisation discussed in this section are skewed in favour of radicalisation, as there are more publicly available sources on this subject than on deradicalisation. As radicalisation concepts fail to adequately explain why some people do or do not become violent extremists, deradicalisation as a concept is also thematically deficient. If the 'symbiote' source of the latter is conceptually weak, extended critical analysis becomes an irrelevant exercise in this regard. In other words, if radicalisation is a weak concept, then deradicalisation is pointless.

Radicalisation and Psychology-Based Concepts Within P/CVE

According to Francis Fukuyama, 'ideas' (radical or new) lack power unless they address the concerns of a large number of people, and those ideologies (extreme variants) cannot be determined solely by social conditions and their associated forces [38]. This perspective lends itself to one central theme in radicalisation discourse: the fact that radicalisation discourse has two distinct outcome effects. The first is that to 'make sense' of the 'new' terrorist threat that has emerged over the last two decades, a process of 'sense alignment' at the political and academic levels has been required. The threat must be perceived as 'new' [41]. The second effect occurs when a 'new' concept addresses the concerns of large groups of people (politicians/governments in general and 'mainstream' society). The net effect of the radicalisation discourse's creation was intellectual agency [65]. This manifested itself in academic and policy circles, which is understandable. However, it also fuelled the 'second-level' narratives that ultimately determine the 'net worth' of any such lines of effort when considered by mass media, fringe media, anti-establishment groups, and conspiracy-based networks [85]. Once developed, the concept became a vehicle for frequently brusque assumptions from those studying it, and it also weaponised 'othering' on a scale that VEOs could not; a development that has become synonymous with the rise of a post-truth approach to information and knowledge, which has morphed into alternative facts and conjecture.

Radicalisation (the concept) exacerbated the fear and anxiety generated by the post-9/11 climate [41]. It is beyond the scope of this literature review to determine whether this was deliberate or an unintended consequence. There is a more straightforward reason for this manifestation: a concept cannot exist independently of the environment it seeks to comprehend. As a result, it becomes a part of the operating environment, providing both benefits and drawbacks, as most things do [49]. This echoes Petley and Richardson's argument that P/CVE efforts must consider the impact on the 'communities' they seek to address, protect, or engage. As a concept, radicalisation has grown too voluminous for any central theme to emerge, particularly in the decade following 9/11. This has changed somewhat, but it is critical to trace the concept's origins to determine how it can best serve strategic communications efforts in the future.

Randy Borum emphasises that radicalisation alone cannot adequately explain why people resort to violent means to meet personal or group needs. Borum emphasises that the perceived causal link between radicalisation and terrorism is critical and that it is the 'involvement in terrorism' aspect of radicalisation that propels it into the

CT spotlight [16]. By viewing radicalisation as a 'collection of distinct processes', this investigation hopes to shed light on the terminology and its application in CT efforts. Borum makes it abundantly clear that radicalisation's conceptual significance is tied to those actions 'that not only justify but also compel violence, and how they progress—or do not progress—from thought to action' [16]. The emphasis here is on the causal relationship between what and how people think and act [16]. This demonstrates that various personal and broader functions may be at work at various levels and that any concept of radicalisation must aim to capture these while also accepting that 'one size does not fit all' [16]. The issue of seeking universal applicability of the concept arises, seemingly undermining the discourse's overall credibility.

Borum's analysis serves as an excellent springboard for more recent perspectives on radicalisation discourse and the theories that underpin the diverse viewpoints advanced by various academic paradigms. Kurt Braddock's extensive research reveals that the conceptual element of radicalisation has been approached and understood in four distinct ways [18]. He believes these four approaches are identity negotiation, depluralisation, self- and other-deindividuation, and dehumanisation. These four trajectories are self-explanatory in that identity negotiation is a natural part of the human condition, as is depluralisation, the process by which individuals begin to lose external affiliations or connections and develop more group-based beliefs due to shedding their previously accepted identities. This process of deindividuation is also connected to how these people apply the same to others outside the group, in that these outsiders are no longer considered 'individuals' but rather members of the 'out-group'.

This process increases the likelihood that these individuals will gradually become more receptive to the idea that outsiders are inferior to them, subhuman, and deserving of derogatory labels such as vermin or animals. Ironically, these four distinct elements fit together well. Still, they do not address how individuals radicalise in terms of the process by which individuals shed their identities and adopt group beliefs [18]. Some of the answers to this 'how' question may be found in the 'assimilation of motivational knowledge' approach. This perspective seeks to understand how the group imparts elements of ideological knowledge to recruits to foster and develop loyalty, a sense of duty, and, when necessary, technical skills (bomb-making, for example) [18].

Bruce Hoffman, a veteran of terrorism studies, has long asserted that the VEOs' lifeblood is their ability to transfer motivational knowledge from adherent to adherent in a cycle of transference that benefits the group's long-term survival needs. Hoffman's assertion provides valuable insight into how beliefs and motivational knowledge are transferred and maintained [58]. However, despite being exposed to narrative elements in various standard processes and interactions, it does not address why some people accept the ideology while others do not. This demonstrates that motivational knowledge alone does not ensure an individual's acceptance or adherence to extremist ideology [18]. Marc Sageman, Thomas Hegghammer, and others have argued that network affiliation is critical in sustaining loyalty and eventual action [55, 106]. This view has three crucial components: the development of social affiliations with extremist movements through peer engagement, kinship, or religious connections; the advancement of one's personal views through engagements with like-minded others; and the acceptance of the movement through social connections with individuals who may have been the original initiators of discussions about the ideology or movement [18, 55, 106]. The social affiliation is also connected, in a small way, to McCauley and Moskalenko's concepts of radicalisation. Still, the outcomes of their respective models differ in that they initially examined this from a linear perspective. The emergence of diverse radicalisation models offers support for a heterogeneity argument alongside suggestions that the concept is too abstract to be of overall use in a deployment sense (CNs).

Radicalisation and Pathway Process-Based Approaches

McCauley and Moskalenko drew on social psychology to develop twelve 'mechanisms' that serve as gateways to terrorism. They refer to 'mechanisms' as 'the means or manner in which something is accomplished.' Thus, vision's mechanism encompasses both the physical stimulus and the associated physiological and neural processes [82]. Individual, group, and mass radicalisation are subdivided into three categories using a pyramid model to identify cognitive and behavioural aspects of each mechanism that combine to form a comprehensive outline of various stages and forces that are both pre-existing and developed during the process [82]. Within these sub-categories, an element of interplay helps explain the complex relationship between self-perpetuating factors and the emergence of group-think contexts.

The first level(s) cite the existence of personal victimisation as a source of grievances that frequently manifest in the realm of political thought and action. These two layers (political thought and action) then play into the hands of a group-based or social interaction setting, which has two additional entry points defined as the 'slippery slope' and the 'power of love'. The possibility of joining a group as a result of pre-existing attachments (relationships) or the formation of such bonds within the group setting contributes to the possibility of the slippery slope, which can be thought of as the gradual elevation of risk factors [82].

This emergence of the 'risk' element is intended to account for the growing academic interest in the concept of 'risky shift' behaviour, which these two authors refer to as 'extremity shift'. As we progress into group dynamics, risky shifts are exacerbated by increasing levels of cohesion, isolation, and perceived or real threat. The concept here is that people are willing to take greater risks to address impending choices once they are fixed in a group setting that limits their ability to make rational choices. Instead, they focus on aims and objectives that may appear to be outside the control of the individuals pursuing them. This has been demonstrated to varying degrees within terrorist groups, and it may apply to the behaviour of the 21/7 (July 2005 terror attacks on the London transport network, two weeks after the 7/7 attacks). Al-Qaeda plot members appeared to exhibit the hallmarks of increasing tension alongside a growing desire to 'act' despite the risk being expanded beyond the cognitive capacity of each individual's rational choice-making psyche. The decision-making phase of the 'failed' operation was hampered and became more desperate in some way because of increased law enforcement efforts in the aftermath of 7/7.

The mechanisms then progress to a stage of comprehension in which the internal and external forces affecting the group's behaviour are captured. According to traditional economic concepts of competition via market forces, the group must compete for support within a single 'market' or diaspora with other suitors. Internal competition for status exacerbates this, as evidenced by the formation of splinter groups and emerging leaders willing to form new groups. The Provisional Irish Republican Army and Al-Qaeda are excellent examples of this, with several offshoot hybrids forming over time as a result of internal and external forces impeding sustained and cohesive singular group structures [22, 82, 106, 112]. A critical point to make here is to contrast this model with whether certain groups intend to form 'new groups' to survive and adapt to emerging external threats from state responses to terrorism. There is a connection here to whether terrorist networks are centralised or decentralised and whether they are created by chance or design [58, 106].

Moving on to the final few mechanisms, McCauley and Moskalenko discuss how increasing pressure for cohesion through 'fissioning' is frequently a failed policy due to the creation of intense rivalry and disagreement. Following that, a slightly different level explains how mass radicalisation can affect out-group behaviour in the aftermath of terrorist acts. 9/11 serves as an excellent illustration of how sentiments of nationalism and patriotism can supplant the cohesion elements of small groups. Indeed, this point can be made equally for victims of state terrorism and for victims of the 'war on terror' who were non-combatants but went on to develop this dynamic of mass radicalisation. As such, are the examples of Iraq, Afghanistan, and, to a lesser extent, Pakistan also examples of how this mass radicalisation can manifest itself in a variety of ways? [122]. The final mechanism summarises this process succinctly by connecting out-group conflict to emerging feelings and sentiments of hatred, thus completing a cycle that begins with individual needs and ends with group-based hatred and conflict as a solution to these needs [82].

The problem with this approach and Moghaddam's 'staircase model' is that it ignores the individual's life events and emotional ups and downs that are part of everyday life for moderates and extremists alike. The McCauley and Moskalenko model was later refined to create the 'double pyramid model', which attempted to distinguish psychological radicalisation from violent behaviour. The central tenet of this new model is that individuals can easily be assigned to 'tangents' of both the values or opinions and action pyramids. The model considers the concepts surrounding sympathisers, fence-sitters, neutrals, and those who feel 'morally' obligated to act [18]. This more balanced perspective emphasises that an individual may be affected by various factors at any given time. There are, however, additional considerations [90, 91].

Other branches of social science have contributed more individualised perspectives on radicalisation processes. Rambo takes a more integrated approach, identifying seven components that he believes have cumulative and recursive effects on one another. Rambo begins with a setting that leads to a crisis that initiates a quest. This quest then results in an encounter, which allows for interaction, leading to commitment and, eventually, some form of consequence [104]. Borum asserts that 'this appears to be quite consistent with observations of post-2001 militant Islamist radicalisations as well' [14, 15, 18]. Despite appearing to represent these movements accurately, there is an apparent lack of empirical or contextual evidence to support this model.

John Horgan examines how cognitive behavioural therapy (CBT) can provide discourse with three 'situational' factors that can influence various behavioural outcomes: personal factors, setting events, and social/political/organisational factors [59, 60]. This perspective's strength lies in its use of the term 'personal', which can refer to any experience encountered during the radicalisation process or pathway. This is balanced by the situational realities people face and their impact on decisions, leading to the social/political dimension, which is the ultimate manifestation of how an individual adopts or believes in a particular ideological standpoint. Additionally, Horgan and Taylor argue that radicalisation is a complex transactional process, emphasising the role of the extremist group in the process as a whole. They regard the way a group communicates its ideology as a critical factor in determining the angles and trajectories taken by individuals [62].

This is an appropriate point to direct the review's attention to classic theories of new religious movements and their emphasis on psychopathological understandings. These theories are characterised by the underlying assumption that an abnormality must exist for an individual to enter a radical group and commit a violent attack [112]. Historically, this approach has relied on secondary and anecdotal evidence to substantiate its central thesis [112]. Additionally, this model is opposed to Crenshaw's rational choice model of the terrorist as a 'rational actor' and demonstrates a shift in the understanding of terrorism studies in light of new global threats [58].

A portion of this emerging understanding also considers the concepts and issues surrounding self-radicalisation. Borum, on the other hand, asserts that this discourse remains rooted in the traditional conversion theory understanding of 'conversion motifs' [16]. Individual encounters resulting from reading, attending talks, and other less personalised exchanges can influence an individual to self-radicalisation before a formal encounter or group setting is created or required. At the moment of interaction, the process can expedite the individual's readiness for deployment based on the preceding process they have already completed. This indicates a solid connection to the current risk posed by 'lone wolf' radicalisation, which is heavily influenced by Al-Qaeda narratives in the online domain [6].

Continuing the discussion of the conversion motif-based approach to the radicalisation discourse, Lofland and Skonovd identify five such elements that exhibit a distinct lexical approach directed exclusively at the religiously motivated conversion context. They begin with an epiphany-based motif termed 'mystical', followed by an 'experimental' process of seeking connections. The third stage of conversion involves the development of solid bonds or sentiments termed 'affectional'. At this point, the group experience occurs, which Lofland and Skonovd classify as a 'revivalist' conversion motif, followed by the 'coercive' element of brainwashing [76].

Borum's analysis of this model indicates that 'while mystical and revivalist motifs have less resonance with the radicalisation experience, experimental and affectional motifs are prevalent among individuals known to have become radicalised into violent extremism, particularly militant Islamist movements and terrorist cells' [16]. Borum is confident that, despite its age, this model still provides one of the best 'conceptual utilities in radicalisation discourse.' Borum concludes by stating that three critical lessons emerge as a result. Firstly, radicalisation must be viewed as a process. Secondly, he is enthusiastic about the conversion motif-based model's potential for identifying a diverse range of radicalisation pathways. Thirdly, Borum views recursive understanding as an attempt to de-emphasise the dualistic nature of the discourse, concerning predisposed and situational conditions [16]. Thus, the discourse's general framework becomes one in which 'the majority of theories on radicalisation focus on at least three factors: (1) developing antipathy towards a target group; (2) developing justifications and mandates for violence; (3) removing social and psychological impediments to violent action.' The development of psychological models of radicalisation processes points to another critical issue with the discourse. The sheer number and scale of these models, whilst suggesting a heterogeneous approach to the field, also exacerbate issues with the terminology and lexicons used in and around academic, practitioner, and policy spheres. It can be argued that these issues confuse the applicability of radicalisation between the conceptual angles of each iteration and their real-world usage in law enforcement, community engagement, CNs, and policy frameworks. As such, a brief discussion on this complexity is essential to bridge gaps in current understanding.

Issues with Radicalisation Terminology and Usage in P/CVE

One of the most challenging issues revealed by this examination of radicalisation concepts is the possible misappropriation or misuse of such terminology versus the terminology simply being 'confusing'. Goli and Rezaei, for example, conducted an empirical study in 2009 using quantitative methods (telephone surveys) and a stratified sample of 1,113 Muslims in Denmark [50]. Their objective was to determine what it means to be a 'radical Muslim' in the context of CT. Their findings indicated that 70% of the sample's 'radical' group were 'predominantly Sunni'. Sunni is an overly broad term that does not provide a concrete basis for defining radical ideology and, according to Githens-Mazer, does not apply to what has been dubbed 'Takfiri Jihadi' ideology [47]. Such use of language and terminology could cause weaknesses in research and insight parameters or be based on predisposed modes of confirmation bias regarding the issue being studied among the researchers. For example, taking this line of logic and terminology further, if Sunni were a proxy for potential radicalism, 80% of the Muslim world should fall into this category, based on Sunni-Shiite demographics over the last century [2, 50]. Being 'Sunni', in the traditional sense of the term from the Islamic perspective, predisposes one to follow one of the four schools of jurisprudence (Hanafi, Shafi, Maliki, Hanbali), which runs counter to the central tenets of 'Wahhabi' or 'Salafist' doctrine, which emphasises a literalist interpretation of texts without the need for 'Taqleed' (blind allegiance) to any school of jurisprudential thought [2].

> If one considers Peter Neumann's blunt proposition:
>
> "radicalisation is what occurs before the bomb explodes". [94]
>
> This can be contrasted with Randy Borum's assertion:
>
> "no single theory is likely to account for all violent radicalisation". [16]

The underlying narrative here appears to be that using this term and its auspices in global P/CVE efforts is, to put it mildly, suspect. Equally important is the foundation for understanding how this concept explains or does not explain the current ideological and operational threat emanating from what remains of Daesh's emergence and longevity. Does radicalisation explain longevity in how extremism evolves and navigates change? This analysis will be presented later in the research study's main findings. It is necessary to discuss the issues surrounding the use of the term deradicalisation in the P/CVE context as a natural by-product of radicalisation discourse.

Deradicalisation

Considerations plaguing radicalisation discourse can be easily applied to deradicalisation debates and hypotheses. There is little doubt that the concept of radicalisation developed more as a means of framing 'new terrorism' in 'new' ways, and this has resulted in the emergence of a bubble of conceptual ideas with little or no long-term robust efficacy beyond policy circles [7]. When applied outside policy departments, think tanks, and university departments, the concept is easily dismantled for its reliance on factors, indicators, profiles, psychological assertions about mental health, and even Islamophobic overtones that are either too specific or too broad, based on assumptions with little or no conceptual weight, and frequently confound an issue that is more easily described as socialisation or, if necessary, socialisation into persuasion [18]. When this is applied to deradicalisation, additional complications arise. At its core, this concept implies that someone who has been radicalised can be deradicalised, indicating that the process by which they became radicalised can be reversed. The narrative conceptualisation of radicalisation and deradicalisation has a fundamental flaw.

Both concepts lack sufficient empirical evidence to establish the 'relative weighting' accorded to individual choices and decisions to enter and exit the process [65]. This flaw exists because both concepts were 'created' to meet the needs of P/CVE efforts rather than being the result of long-term observational and quantitative/

qualitative research. If, for example, as Francis and Furedi assert, radicalisation is no longer seen as a social process (as would be the case due to its over-emphasis on the psychological and cognitive elements), the current academic study cannot confidently assert that it has identified the genuine 'critical factor' in an individual's decision to adopt extremist thinking. This is also true for deradicalisation, as it can be argued that this process is a form of re-socialisation back to the individual's previously known social reality.

The P/CVE community has established distinct models that provide a set of parameters. However, these parameters paradoxically lack the understanding and tools needed to produce anything more insightful than a wide range of potential indicators, factors, pathways, processes, and profiles. Consequently, radicalisation and deradicalization emerge as highly contentious concepts. Nonetheless, there are potential solutions to address these shortcomings. In terms of deradicalisation, it is essential to consider other 'outlier' factors that may influence how the process is perceived and related to sub-elements such as disengagement, desistance, and recidivism. The involvement of former extremists in deradicalisation efforts is central to this discussion. These sub-contexts may be regarded as 'outlier' discourses within the field, yet they are important components in understanding the evolution of both theoretical and practical aspects of P/CVE discourse.

Outlier Discourses Within Deradicalisation Studies

According to the testimony of former extremists and violent extremists [100–102], two critical processes are at work at every level of engagement, i.e., doubt, certainty, acceptance, and rejection. The first is a process of socialisation; the second is a process of persuasion. The socialisation component may be familiar to the individual or unfamiliar, depending on the nature of the individual's context. The processes of familiar socialisation are those that are based on 'mainstream' norms and values. Less familiar are the processes that build on this initial component by amplifying, enhancing, and empowering ideas, narratives, and emotional triggers. If socialisation, as a concept and observed reality through social science, has any conceptual value over radicalisation and deradicalisation, it is because it avoids fixing entry and exit points. This allows for consideration of the full spectrum of cumulative life experience without the forced rigidity of radicalisation and deradicalisation concepts. When the issue of persuasion is added to this mix, a succinct but multi-layered approach to comprehending, explaining, and addressing the issue of extremist recruitment, retention, and potential rejection of extremism can be advanced (disengagement, etc.).

Deradicalisation and a Perceived Lack of Clarity

Unsurprisingly, the ambiguity surrounding radicalisation and its application makes a detailed discussion of deradicalisation somewhat tricky. However, several critical points identified in this research study require vital examination, if not critical analysis. According to available sources, deradicalisation is primarily viewed as a process of abstaining from violent means, accepting the broad norms and values of mainstream society, and, in the Western context, recognising that social or political change can only occur through the application of democratic principles [7, 18, 115]. The issue identified in this study is that radicalisation is such a contentious concept that, aside from disengagement concepts, deradicalisation remains a 'work in progress'. Disengagement from terrorism is qualitatively distinct from deradicalisation. The former merely entails a decision to abstain from violence at a particular time. The latter appears to be a process of reintegration into the very milieu that many argue contains the drivers of radicalisation in the first place [47].

Data sets from deradicalisation-specific initiatives still need to be fully public in a standardised, research-specific paradigm of uniformity. This makes it very difficult to determine whether the concept is appropriately understood. More importantly, it raises the question of whether disengagement and deradicalisation are as susceptible to predisposed and situational factors as radicalisation appears to be. In other words, if extremists employ obscure and nuanced theological or ideological concepts to conceal future involvement in violent action, and this individual has been exposed to the core Jihadi narrative, how can it be asserted that they have ceased specific extremist behaviour through disengagement or have been deradicalised? This is an admittedly blunt statement, but it is intended to demonstrate that, in the absence of contextual and conceptual clarity, these terms hold only lexical meanings in the context of the current 'state of the art' in P/CVE. It also demonstrates, hopefully, that by viewing deradicalisation as a process of re-adopting previously 'rejected' values, the likelihood of these values being compatible with the expectations of liberal Western democracies is a somewhat difficult model to pursue at this stage of P/CVE understanding. When one considers Daesh's current 'unknown' shape and framework in the West following its 'defeat', such a pursuit becomes even more shrouded in 'known unknowns'. The following statement locates these points differently but conveys the same message intrinsically when it states that 'as long as there are those seeking answers, a true identity, and perhaps a sense of adventure, there will be a market for Jihadi radicalisation in the West' [1].

There appears to be some recognition within the P/CVE discourse of the relatively straightforward identification of identity, belonging, and culture as critical aspects of radicalisation and deradicalisation. However, this study discovered that much of this is motivated by a view of these issues as influenced by heritage and cultural forces on the one hand and by the pressure to conform to Western versions.

When disengagement, desistance, and recidivism are incorporated into the deradicalisation discourse, the 'causal variability' issue become so significant that each process requires its own distinct conceptual framework. They do not all fit neatly into the deradicalisation box. This study proposes a new conceptual framework based on 'communicative socialisation' as a prelude to refining radicalisation discourse through communicative means at the sub-level of CN work within P/CVE. The goal is to begin eliminating these discrepancies to arrive at a more streamlined approach to extremism and violent extremism, unafraid to acknowledge ideological, policy, and conceptual development issues by accepting and understanding the strategic roles of socialisation, communication, and persuasion. This approach could be applied through the lens of national security, foreign policy, or integration. This effort would assist the discourse in moving beyond the accusations that it is anti-Muslim, contains prejudicial undertones, or is fuelled by confirmation bias [53, 107].

This review has so far reflected some of the disparity found within central and more outlier radicalisation and deradicalisation discourses. The broader argument could be made that national security 'concerns' have frequently been used as a vehicle for focusing on radicalisation rather than deradicalisation, as the former is viewed as an external process towards criminality rather than the perceived absence of justice that the 'radical'/extremist narrative espouses. Acknowledging justice's 'perceived' absence would be a complex political and policy stance. Numerous academics critical of the P/CVE discourse point to this rather significant anomaly, and the overall opinion appears to hold some gravitas. Alternatively, one could argue that the looming shadow of state-funded research into radicalisation and deradicalisation highlights grievances as always being 'perceived' rather than 'real'. Is it not true that grievances are always 'perceived' in some way, and that this perception does not make them any less 'real'? Much of the research reviewed appears to overlook the legacy impact of the war on terror and the foreign policy.

Reviewing radicalisation and deradicalisation concepts has revealed that radicalisation and deradicalisation frequently have a paradoxical relationship within P/CVE approaches. These are commonly defined by competing needs and opposing expectations. Such disparate needs are then exacerbated by a paradigm that conceptualises radicalisation and deradicalisation as 'absolute' concepts, rather than seeing them as relative terms that require contextual clarity to address specific threats, trends, and developments [109]. The criticism stemming from the confusion levelled at state-sponsored research findings aligns some academic efforts to affirm emerging or, at best, embryonic trends and risks (such as Daesh 2.0), creating tangible cause for concern in this field. This leads to further segmentation of these subjects within foreign policy, security, and integration, resulting in either broad parameters or undesirable or harmful outcomes [109]. The disadvantages include focusing exclusively on pathways to radicalisation and ignoring the circumstances and factors that, one could argue, are just as powerful triggers of extremism as any ideological indoctrination. It appears radicalisation and deradicalisation are conceptually ambiguous, too broad, or too specific in some cases, and deficient in application. They also suffer from a 'communication deficit' in how both discourses are used within official circles and how they have been presented to communities. There is still much work to be done to bridge gaps in understanding and application. This suggests that socialisation and re-socialisation approaches may better suit the P/CVE sector's needs and allow for both adaptive and more comprehensive conceptual ideas to come forward.

In comparison, the preceding sections point to the need for further review of subsections within socialisation theory to be discussed. The key elements within this are communication and behaviour change issues. Without communication that seeks to alter, modify, or change behaviour, socialisation theory is useless as an umbrella for how norms and values are acquired, accepted, rejected, and adapted.

Strategic Communications Studies

From a logical framework standpoint, strategic communications within CT are predicated on a policy and strategy foundation to handle crisis response narratives and counter-extremism activities aimed at specific VEOs or violent extremism in general [69]. The scope of this research is narrowed to those directly generated and managed to oppose VEO narratives in the digital arena. This is for two reasons: first, the emergence of extremist narratives on the internet necessitates an in-depth examination of CN initiatives. This enables comparative analysis, as well as conceptual and empirical research, and evaluation of the political, social, moral/ethical, and political/social settings in which these tales exist. Second, counter-extremist communication enables actors and scholars alike to assess and evaluate alternative approaches, which can aid in formulating policy and strategy for CT operations in general. As such, this emphasis reflects a 'bottom-up' approach to studying counter-extremism communication while providing a 'funnel' picture of how CT policy flows through academia, practitioners, and the tech milieu to reach audiences. This work enables a conceptual and technical examination of the efficacy of CT policy at its 'sharp end': the point at which judgments, choices, and impacts occur in the public arena.

In recent years, the discourse on terrorism studies has sought to integrate disparate academic perspectives into a more multidisciplinary approach to specific epistemological and ontological concerns, such as what causes terrorism. How can someone become involved in extremism, and conversely, how can they be persuaded of it? Despite this gradual fusion of perspectives, the discourse is still dominated by the application and underlying knowledge of psychological and criminological narratives and understandings from a theoretical perspective [62, 65, 92, 113]. Frequently, the synthesis of these two methods has resulted in a multidisciplinary discourse on terrorism studies. From Crenshaw's idea of the rational actor to Horgan's environmental psychology and cognitive studies approach, the discourse has recently begun to dissect and understand the function of narratives, communication, prevention, and audience engagement in the online or digital realm [18].

Rapid and revolutionary changes have occurred in the technological aspects of modern communication and virtual community composition. These seismic shifts in political and social mobility narratives and structures have created a significant knowledge gap. Additionally, there needs to be greater contextual clarity regarding how contemporary practices in online P/CVE work and their effects. There are few if any, theoretically-oriented narratives on which to base analyses and evaluations of the digital age of strategic communications within P/CVE [46, 64]. This section will present essential theoretical concepts that can be used to test and examine assumptions regarding the role of CN communications and provide a solid foundation for future CN efforts. The most well-developed model in this guise is the RAM (illustrated below), which will form the operational elements of a socialisation and communications-based model for this study. The RAM approach will be discussed in relation to its merits later in this study. Counterspeech is broadly understood to denote: "any direct response to hateful or harmful speech which seeks to undermine it" (Dangerous Speech Project in [117]).

A counternarrative (CN) or counter speech effort can manifest in diverse ways, such as communicating to large audiences, specific audiences, one-to-one, one-to-many, many-to-one, and many-to-many [117]. The case study analyses (Chaps. 4 and 5) seek to better understand elements of these approaches in detail for their perceived efficacy and outcomes using the SoRaSCo model. This study will use counternarrative (CN) instead of counterspeech where relevant for ease of access and understanding (Fig. 2.2).

Any attempt to theorise communication must start by differentiating between the two main theoretical approaches to the subject matter. These can be categorised as the transmission model of communication and the ritual model of communication. Seminal thinker John Dewey championed the former in his book 'Experience and Education', and in the broadest sense, suggests that communication be viewed as the dissemination of knowledge and not just 'expression' per se [30]. The latter approach

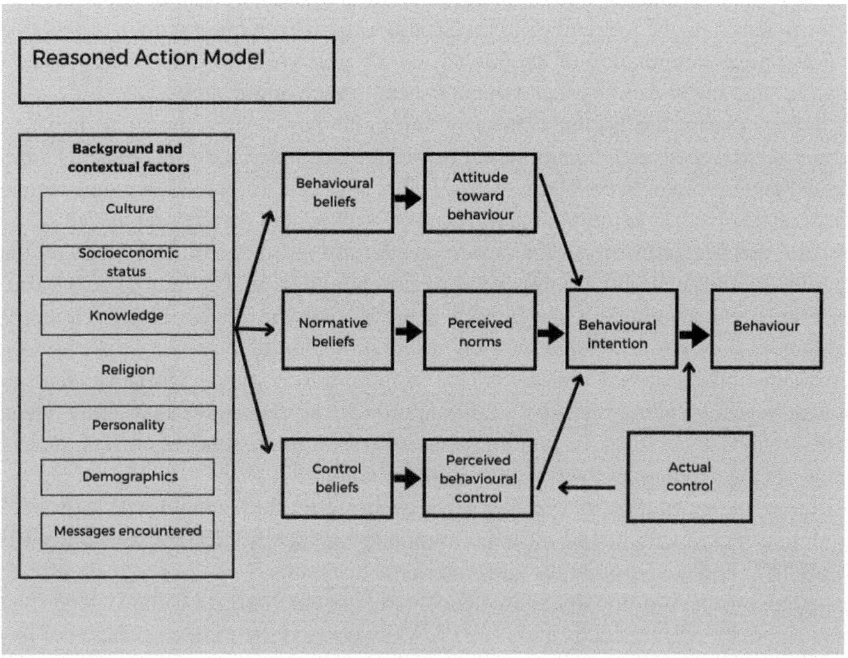

Fig. 2.2 The Reasoned Action Model (RAM)

is most associated with another giant of this field, James Carey, who famously said 'life is a conversation' [23]. This was suggestive for Carey of the view that, for social orders to be critically understood, the first task is to reflect seriously on the capacity of humans to think and to replicate or fabricate relevant symbols through which a shared symbolic order emerges. What Carey was doing, in essence, was relocating communication to a position of primacy concerning how society functions and how human beings learn to not only function but make active meanings through communication and interaction [24]. As pointed out by other critical thinkers such as Raymond Williams, such an approach leads to a broader understanding of the developmental cycle of Western society, in what Williams coined as the 'long revolution' of communication.

Strategic Communications as an Extension of Communication

As would be expected to some extent, strategic communication is a multi-disciplinary area that has relevance to many sectors of later modern or post-modern society. Strategic communications are the 'purposeful use of communication by an organisation to fulfil its mission' [51]. It can be argued with significant validity that this field is as old as the organisations that use it, since they always intrinsically have goals and communicative entities [119]. The diverse fields in which strategic communications are applied mean that some effort at bridging gaps between traditional understandings of public relations, marketing communications, and organisational communications is one of the concept's key strengths [51]. This study focuses on public authorities, interest groups, civil society organisations (CSOs), and private companies. All these are classified as organisations of some sort, so they can be looked at from a strategic communications angle regarding how they use this tool for their various objectives.

From an organisational perspective, strategic communications can be seen as conscious organisational communication efforts to reach the overall goals [34]. Therefore, the organisation's overarching goal(s) would be at the centre of this effort, not the communication effort's sub-goals. It is about long-term identified targets expressed and articulated through top-level communication. Strategic communications, therefore, are based on an awareness of the essential importance of the use of communication for the existence and performance of the organisation [51]. In the context of P/CVE communications, this may be a case of the 'effort' being part of social moral responsibility and the maintenance of the authority, credibility, or legitimacy of public organisations and CSOs. In many respects, much of the theorising about strategic communications does not neatly fit into the P/CVE context, which is somewhat surprising. Nevertheless, all is not lost in sourcing clear conceptual foundations for this project. This is because, at the heart of P/CVE, CNs', communication is more distilled in its essence before words like strategic are added. It has to be remembered that for the end-user, it is communication, whereas for the sender, it is strategic in focus. Therefore, we can unravel the strategic communications process in P/CVE by breaking down its constituent parts for a clearer understanding. We start with communication before moving on to what strategic may or may not mean in this context to aid this study's theoretical approach.

Definitional approaches to communication seek answers and direction. Specifically, they aim to understand what communication is and what about the nature of communication makes it an intrinsic part of human development. With such a vast and often complex paradigm, it is no real surprise to find that, from a definitional perspective, there are extensive iterations of what communication entails. When these are specified in some way, they still come across as attempts to be 'catch-all' definitions. In Shannon and Weaver's definition, we see perhaps the broadest of all parameters in this regard: "communication is all of the procedures by which one mind can affect another" [110]. Anderson, however, according to Cox, introduces related elements that form part of the communication process as a whole: "communication is the process by which we understand others and in turn endeavour to be understood by them. It is dynamic, constantly changing and shifting in response to the total situation" [26].

What is interesting about this definition is the use of the terms process, understanding, and dynamic. These terms imply meaning, interpretation, and interaction as part of the communication process. If we look at how Miller has defined what communication is, we again find ourselves in a broad area of view that does not provide any line of focus or specificity; "communication means that information is passed from one place to another" [89]. Although rudimentary, this definition does provide the most basic and distilled version of the act of communication, but not the art per se. One easy comparison here is with the continued difficulty in finding a universally accepted definition of terrorism, which also goes some way to explain why the overall discourse is so convoluted.

The idea of communication as a process was developed further as the influence of technological advancement started to move beyond traditional understandings of communication. Carey famously pointed to the use of the telegraph and the advent of the printing press as ways in which communication evolved and became more about processes, symbols, and meanings that took on frames of their own within social development and norm creation as industrialised society started to boom [23]. Perhaps the most intriguing and relevant aspect of communication to online P/CVE efforts can be found by understanding what Berlo means by stating:

> "...when we label something as a process, we also mean that it does not have a beginning, an end, or a fixed sequence of events. It is not static or at rest. It is moving. The ingredients within a process interact; each affects all the others" [12].

This stance becomes important for CN efforts because it implies that such work often relies on time-specific dissemination and does not consider longitudinal effects. This extends to being unaware of unintended consequences and transactional effects from actors (audiences) for whom the original communication was not intended [64]. When we see communication as a process, we can understand how even simple exchanges are influenced in intricate ways by the past and how this will have significant implications for the future. This process was linear in early conceptualisations of communication, with communication moving from a source to a recipient. This is seen in Laswell's classic communication model, which regards the fundamentals of this field as being drilled down into: "who says what, to whom, through what.

The Development of Communications Studies

The linear approach gave way to what is now known as the transactional communication model. The most pertinent aspect of this approach is found in the difference between 'action' (linear communication), 'interaction' (engagement) and 'transaction' (meanings). This suggests that the reactions of various types of receivers warrant equal attention concerning how we measure and monitor the efficacy of source communication from the sender. In theory and practice, a person acts simultaneously as sender, receiver and responder [21]. From a theoretical and research perspective, this somewhat conveniently leads to further depth being added to communication studies as a whole from the sociological study of semiotics. Developed originally by Ludwig Wittgenstein, semiotics seeks to understand the role of symbolism in the creation of meanings through various uses of what are known as signs and their associated signifiers [25, 77].

Symbolism and Meanings in Communications

Symbols, semioticians argue, carry no natural or innate meaning, as they hold only arbitrary relationships to what is being symbolised. Meaning is subsequently created in them through what they become attached to via language, culture, and custom. These meanings, then, are shared in a 'lived social experience' and are accepted as the expression of the various symbols prevalent in any society at any given time. Through the use of these symbols, interactants create meaning, both actively through exchange and passively through norms and values, which incorporates socialisation processes [36]. When theorists claim that communication is symbolic, they suggest it necessitates using signs and symbols that have arbitrary links to referents. This means that in terms of communication theory and its development, slowly but surely, a process of evolution in understanding has resulted in two main divergent conceptualisations: communication as social activity and communication as intention.

The first question regarding communication as a social activity would be what constitutes communication. Must at least two people be engaged in the process (interpersonal), or can it also mean that an individual communicates with themselves (intrapersonal)? Scholars differ on this issue, as some believe that intrapersonal communication is not communication per se, but rather the mechanism and vehicle of cognition or thinking, and that communication requires at least two people (Laswell, 1958, cited in [88]). Arguably, however, the more pertinent issue is that communication serves as a social vehicle and, therefore, as a fundamental element of the socialisation process. It can be surmised, then, that what communication 'does' in this process between two or more people is of great importance.

Regarding viewing communication from the lens of intention, the central proposition is that communication, by default, carries inherent meaning. The premise is that all 'acts', including communicative ones, are manifestations of human behaviour and therefore hold and contain implied meanings. This relates the process and function of communication to human behaviour. The dilemma here is that one approach regards the presence of intention as the key to the communicative method; in other words, it is not communication if it is not intended, while the alternate approach regards human behaviour itself as a form of communication (Laswell, 1958, in [88]). What this debate does for the current project's research criteria is to offer a key area of concern and enquiry regarding how we jump from communication to strategic communications and why the latter is seen as the domain most relevant to CN efforts within P/CVE.

A picture emerges of why strategic forms of communication appear to have a clear distinction between themselves and a traditional understanding of the nature of communication more generally. For this explanation, offering a working hypothesis on the meaning behind using the term 'strategic communications' in online P/CVE efforts is helpful. The term 'strategic' implies that a set of intentions, aims, and goals has been considered, proposed, and rationalised as viable means to disseminate coherent messages through online mediums to specified target audiences. There is a lot to unpack in this working definition. Still, it can be simplified by adopting the following: 'P/CVE strategic communications use aims and objectives to send targeted messages to identified audiences'. At this stage, however, this working definition does not answer the fundamental question of to what end or why this is the meaning of strategic communications in the P/CVE context. To uncover this anomaly, we go back and revisit, first and foremost, the concept of socialisation, which seeks to embed or re-embed the accepted norms and values of the dominant social system; from this, we can infer that the ultimate aim of P/CVE strategic communications in the CN context is behavioural change [46]. This addition to the conceptual framework requires understanding the social-psychology discipline of behavioural studies through a review of the RAM of behaviour change.

The theoretical points discussed in this section have shed light on both the development and the role that types of communication serve in social contexts. The links between the type, messenger, message, mode, and recipient of different forms of communication offer an advantageous framework to support (political) socialisation theory in this regard. The remaining element of this strategic communications review is centred around the role and need for behaviour change to be a central tenet of P/CVE understanding. Given its place as a critical conceptual element within public sector strategic communications, the RAM approach to behaviour change will be reviewed in this context.

The Reasoned Action Model (RAM) of Behaviour Change

The RAM approach (illustrated on page 46) focuses mainly on conceptual constructs concerning individual motivational factors determining the likelihood of performing a specific behaviour [4]. Developed as an overarching paradigm of study to incorporate the theory of reasoned action (TRA) and the theory of planned behaviour (TPB), this approach assumes that it is, in fact, 'behavioural intention' that is the most effective predictor of behaviour. This behavioural intention is measured through the presence of a confident attitude towards certain types of behaviour, which are, in turn, informed by personal and social perceptions of said behaviour. The development of the overall RAM approach stemmed from the initial TRA model, which added perceived control levels over the performance of behaviour from the TPB model. The TPB components were added to the TRA and became known as the RAM, which was created by Fishbein and Ajzen. Weinstein argues that correlational results cannot accurately predict behaviour. Still, studies in the field of health behaviours have indicated that TRA and TPB, respectively, have been able to shift behavioural patterns linked to issues like smoking, drinking, substance misuse, etc. [121]. It is not immediately apparent how such studies link to extremism and counter-radicalisation understanding until the assumptions of harm reduction and cognitive behaviour change models are added to the discussion as the 'industry standard' tools of CT in prevention and safeguarding contexts [35, 62].

Developed in 1991 as an added feature to the original TRA by Icek Ajzen, the RAM was introduced and further developed along the central idea that behaviour is predicated by attitudes and intentions towards the behaviour in question, which can be managed and, therefore, to an extent controlled or planned for. As such, the original TRA eventually became the RAM, incorporating the 'planned behaviour' element. In their own rights, these models are cognition-based theories regarding their etymology. Both models assume that humans can 'apply' reason and that reasoning is the primary psychological process involved in decision-making [75]. According to Dutta-Bergman, one of the major criticisms of the TRA (which, technically, also applies to the TPB as an extension of the TRA) is that its cognitive solid orientation tends to exclude the affective (mood, emotion, and situational context) nature of humans, which also plays a role in decision-making processes [32]. Using the slightly out-of-context example of sexual behaviour change, the TRA/RAM suggests that for an individual to change his or her behaviour, he or she must systematically identify and weigh the consequences of his or her sexual behaviour to form attitudes toward the specific behaviour that must then be learned.

This assumes that a behaviour change can be induced by adding a new belief, altering the favourability or unfavorability of an existing belief, and altering the strength of the belief associated with the desired behaviour. The same logic is applied to how extremism and extremist ideas manifest and are addressed and managed. Behaviours and attitudes are often very much private sphere-based and, therefore, difficult to predict, similar to latent extremist attitudes, which can live within the emotional private sphere of the individual for any length of time once socialised into the ideology/narrative. This is seemingly why the RAM approach is favoured but open to critique. There appears to be a paradoxical element to predicting behavioural intentions, which often cannot be 'viewed' until it is in the public domain.

The process of persuasion required to change a person's behaviour would be primarily information-based (strategic communications), providing the individual with the narrative or data/information necessary to form a favourable attitude toward the desired behaviour. While the individual may meet the requirements for behaviour change in this paradigm, they may not be able to implement the behaviour if they initiate sexual activity on a whim. This can, to some extent, mirror the sudden process of being socialised into acting out extremist attitudes as a result of some external change in a social context, such as an attack, hate crime, or incident that propels someone to fast-track grievances and other factors into an eventual scenario of acting violently based on an extremist whim or desire. Another criticism is that the TRA emphasises the individual rather than the group to which that individual belongs [32, 95]. Noting the African cultural context in which she is writing, Van Dyk suggests better understanding community involvement in AIDS education, prevention, and counselling [118]. This is especially important for HIV/AIDS research. Van Dyk emphasises the significance of healthcare interventions in traditional African contexts where communal living is the norm. In contrast, the TRA and the TPB emphasise the individual actor and disregard the collective context in which individuals exist. Cultural variance in understandings of what constitutes the 'moderate versus extreme', the role or value assigned to the network versus the individual in radicalisation, self-radicalisation, etc., all appear to be potential independent variables open to wide-ranging interpretations.

Dutta-Bergman suggests that although proponents of the RAM approach may argue that a subjective norm explains the role of the collective in an individual's decision-making, it is still driven by an individual's motive orientation, thereby keeping the locus of decision-making with the individual [32]. Although subjective norm taps into the actor's evaluation of significant others, it does not tap into the complexity of the social fabric comprising the health behaviour in question, which again appears significant for counter-radicalisation/deradicalisation/CN understanding. Dutta-Bergman observes that social influence extends beyond the sphere of a few influential others to the larger socio-cultural context of the community (anti-authority, minority groups). While a person's closest significant others may discourage multiple sex partners, the larger subculture to which he or she belongs (such as the urban youth subculture) may view such behaviour as 'normal'. Due to its emphasis on the individual, the theory lacks a social-ecological perspective that could be used to influence social policy. The social-ecological approach assumes that the community is superior to the individual. As a result, the transformation of individual behaviour results from transformation at the community level.

Susan Kippax argues from a social constructionist perspective (meanings, culture, and processes are social constructions constantly being made and re-made). This extends Bergman-Dutta's argument by emphasising that the TRA overlooks the 'social' or 'collective' nature of sexual (health-related) behaviour by focusing on the individual [71]. They are referring to the social and cultural environment in which sexual behaviour occurs. They suggest that HIV/AIDS research should consider the cultural contexts that shape sexuality and sexual practices, the lifestyle of populations where HIV infections are prevalent, and the interpersonal interactions through which sexual practices are constructed and enacted. According to Kippax, the focus of sexual behaviour research should be on the common sense (or assumed reality) of the populations being considered. Common sense refers to the social discourse or constructed reality regarding sexuality [71].

Relevant to this investigation is the reported prevalence of 'flesh-on-flesh' intercourse among South African mine workers [78]. Whilst accepting this frame of reference can be uncomfortable for the reader, within the scope of a research study such as the one currently being conducted, this type of variance within the behaviour change and social discourse context can be applied to why some who turn to extremism see themselves as tuning not to extremism per se, but to a 'purer', literal, and therefore righteous manifestation of their faith. The social construction of righteousness versus what is seen as deviant would thus manifest through social processes and unique sets of norms and values, i.e. variant socialisation processes. In essence, if the planned behaviour does not correlate with, correspond to, or confirm a pre-held social reality (literalism in faith, for example), there is a risk that the communication attempt will further cement the original mindset and attitude. As such, behaviour change may need to be seen as an amalgamation of discourse and decision-making, wrapped up within elements of which social reality harbours the most significant level and depth of control regarding attitude, intention, and behaviour cues.

A classical approach of using reasoned action/planned behaviour models to change the 'risky' practice of 'flesh-on-flesh' intercourse would entail influencing the attitudes of individual miners so that they view this behaviour as undesirable. It would also seek to influence them to associate with others who hold views 'antithetical' to this behaviour. Individual miners' attitudes and subjective norms would then be aggregated, and an intervention would be designed and implemented to influence their intentions. Kippax would argue that this approach is irrelevant if it fails to address the shared meanings associated with this practice within the mining community. This would entail debating this practice to persuade the community to adopt a different viewpoint. Through these debates, the social representation (the discourse that shapes attitudes) of unprotected sex could emerge. At the level of subjective norms, the intervention would consist of debates and discussions about the cultural values and ideology surrounding unprotected sex, as well as the underlying beliefs that support this behaviour. Individually held normative beliefs (i.e. the belief held by the significant others of individual miners that unprotected sex is acceptable) would need to be correlated with the accepted reality (the broader social construction of sperm accumulation causing mental illness). For the intervention to be deemed successful, the accepted reality of unprotected sexual activity in the community must be altered. Kippax argues that community-level attitudes, subjective norms, and the behaviour they influence are more influential than private beliefs.

Stripping away this dual layer of existing beliefs is perhaps why RAM approaches stress the potency of cognitive intervention. At the same time, they could fail to acknowledge that there can be no cognitive process without behavioural contexts and no behavioural context without social processes. The social exists within the private and public spheres and, therefore, can contain protective layers of normative alignment that a single communicative message cannot access.

These community-level attitudes and norms are social constructions that must be altered to change behaviour successfully. The argument here is that theories of rational action, reasoned action, etc., disregard explanations of social behaviour grounded in the community. These can certainly be argued within the context of extremist beliefs and attitudes that have become interwoven into the fabric of community dynamics, such as conservatively literalist spaces where extremist ideology can take root but manifests only within specific individuals who possess other factorial vulnerabilities that can be exploited or harnessed. The case of Mohammed Sadiq Khan (7/7 attacker) and his counterparts and their social dynamics within the communities of Leeds, Nottingham, and Dewsbury in the UK serves as an example of ingrained and insular conservative and literalist attitudes and lifestyles that were then exploited by nefarious recruiters and ideologues [20]. Kashima et al. present a similar argument to Kippax, stressing the dual-activity nature of partnerships (cells, groups, networks). Kashima and his colleagues believe that the attitudes and norms of the partner(s) must also be measured. Assessing both partners' attitudes, subjective norms, and perceived behavioural control will enhance our understanding of decision-making in cooperative behaviours like condom use.

Individual behaviour is tailored to the causal structure of rational action and planned behaviour theories at a conceptual level, but falls short when other variables are present. It only accommodates cooperative behaviour (7/7 attack and planning), a field of application in which Kashima and his colleagues believe additional research is necessary. In addition, Kippax questions the theory of rational action's linearity. They contend that norms and beliefs do not necessarily precede or determine behaviour or action, particularly concerning sexual behaviour and condom use. The linearity of the TRA is incapable of explaining why risky sexual behaviour occurs despite the actor's correct beliefs and social norms. Again, this criticism can be levied at CNs that attempt to alter behaviour among target audiences where accurate data regarding levels of radicalisation or extremist attitudes have not been collected. Yet, the underlying assumption is that some form of generic vulnerability to extremism must be prevalent. This raises serious questions about such assertions in light of the rise of the decolonised activist and academic movements, through which such assumptions have been epistemologically challenged as well as from moral and structural inequality contexts [120].

According to Kippax, this assumption of causal linearity fails to recognise the complexity and dynamic interaction between practice, beliefs, and normative structures. This also appears to be a key emerging argument for why the RAM approach and P/CVE CNs remain 'counter-intuitive'. This failure is why the RAM cannot explain why individuals exposed to the same situation do not always behave similarly. Again, the issue of radicalisation affecting people in similar social realities differently is paramount and calls into question both the efficacy and conceptual clarity of CNs within P/CVE, which rely heavily on the RAM approach, among other behaviour change models. For Kippax, this proves that the relationship between beliefs, norms, and behaviour is not simple but rather complex and multi-layered [71]. Hobbis and Sutton, in an investigation of the application of cognitive behaviour therapy techniques to behaviour change interventions based on the TPB, criticise the TPB/RAM for failing to guide how to promote behaviour change [57].

The theory is effective at explaining and measuring behaviour, but it is difficult to support the notion that it can also 'predict' behaviour. It does not specify methods for altering behaviour. In the context of P/CVE CNs and their 'positive intervention' premise, this raises a serious question regarding what a 'radicalised' individual should do post-intervention and how this links to 'adjusted' attitudes and belief expression. In other words, how is change, or the lack thereof, measured over time or as a recidivism risk after engaging with a short-term mass communication or targeted CN product? The underlying assumption appears to be that the CN product has 'served' its purpose and cannot do any 'harm'; it only acts as a positive intervention, even over time, and changes social reality. Hobbis and Sutton also criticise the development of the theory's modal (form as opposed to substance) salient belief-based intervention. They object that it is challenging to accurately represent the most critical modal beliefs upon which interventions will be based [57]. In other words, modal beliefs are not uniform to all in the same way and, therefore, cannot be seen as a specific manner or form by which to conduct such an intervention. They argue that an intervention based on modal salient beliefs may be ineffective because many target group members will be exposed to information intended to alter beliefs that are not salient to them.

Furthermore, health management behaviour change models seek to minimise or prevent harm first and foremost to the individual, whereas P/CVE approaches seek to minimise the risk of damage to others from the actions of an individual or group. The individual's vulnerability concerning extremism is primarily a security concern, given the potential social, economic, and security policy impact of a terrorist act. Health behaviour management work has to consider genetic, biological, and physiological variances as uncontrollable variables. In contrast, extremism or propensity to extremism is still without definitional or conceptual consensus or clarity. This anomalous set of issues may also explain why cases of terrorism related to mental illness or mental health-based issues became normative exemplars despite data pointing to mental illness and terrorism links being relatively small overall [67].

Several national counter-extremism and CT policy frameworks were keen to use psychological models to explain why people become radicalised but not necessarily what has radicalised them. Equally, these same policy frameworks have been unable to locate and express what would curtail or affect radicalisation beyond suggesting that mainstream values are essential. Which 'mainstream' values might these be when many Western nations have taken a worrying slant to Far-Right ideological stances and once again amplified populism as a device to address 'grievances' among populations? Political trend analysis suggests that the term 'mainstream' can no longer adequately or accurately capture any significant thread of 'normative' moderation compared to extreme and extremism. Is the new normative narrative of the twenty-first century thus a moderately variant form of extremism, and what does this mean for strategic communications in a prevention space such as P/CVE? Who is moderate is the real mild, and vice-versa, is something that may need to be taken seriously as a question.

Hobbis and Sutton propose an alternative technique for eliciting personally meaningful beliefs. This would result in individually tailored interventions based on the content of each individual's unique set of salient beliefs, rendering group-based interventions ineffective and cumbersome. In seeking an appropriate conceptual framework for P/CVE CNs, the issue of being able to target narratives at an individual level while reaching larger audiences where relevant is precisely where the RAM approach needs significant modification to be effective in this space. The issue with this, however, is that the tech capability to target individuals in this fashion with online CNs is not open to many of the 'credible' messengers unless they partner with tech companies.

Kaiser et al. criticise the RAM's research methodology. According to their analysis, the RAM's behaviour measurement adheres to the compatibility principle. As a means of improving the quality of what a specific concept measures, strict adherence to a measurement paradigm is referred to as the compatibility principle. In the case of the RAM, the principle dictates that this theory must be applied almost exclusively in a specific way to explain a single action. According to Kaiser and colleagues, such specific, single-item assessments lack validity and reliability. Single-item measures focus on a single domain of behaviour, overlooking related behaviours. The use of condoms, for example, is not a singular but an aggregate behaviour [68].

The expression of extremist views cannot be seen solely as a singular behaviour but as the cumulative effect of several interconnected and independent behaviours that merge and manifest through expression or action (why is extremism bad? And why should I change?). The RAM approach requires robust examination in the context of the P/CVE CN space to allow for a full assessment of its potential strengths and limitations. As such, it is necessary to hone the debate towards understanding what P/CVE CNs are and what they seek to achieve. Suppose this is done at a macro level. In that case, the information will allow the case studies-based discussion to drill down into mechanisms, nuances and variances of different CNs and how the learning from these efforts may help inform the future progress of this critical work.

This chapter has highlighted elements of positive and less effective practices within the online P/CVE sphere. In the area of strategic communication and behaviour change, understanding these elements' respective roles within P/CVE CNs, the lack of application of socialisation theory and its facets in the context of the globalised digital world today suggests that behaviour change models, such as the RAM approach, need adaptation and greater nuance.

In many ways, this discussion mirrors those found earlier in this chapter regarding why socialisation, or political socialisation for that matter, are not adequate concepts in their own right to explain what has become known as radicalisation [54]. Suppose it is accepted wisdom that individuals are drawn to extremist narratives through overt and covert forms of new socialisation processes that take them away from existing socialisation processes and their associated values. In that case, this change must include both attitudinal shifts and behavioural changes as key features [62]. In the context of preventing extremism, which is where CNs broadly reside in the overall CT spectrum, the idea of prevention stems from the likelihood of harm to the person or those around them if the behaviour is not challenged through some form of intervention process. It is the aspect of vulnerability that transcends the issues of extremism and radicalisation into the behaviour change arena and places them alongside issues like health behaviour, gang-related behaviours and notions of deviancy developed alongside the concept of socialisation to explain those who fall outside expected behaviours regarding norms and values and the fulfilment of social roles [19, 52].

This translates into a set of assumptions about the nature of radicalisation and vulnerability, as well as cognitive behaviour issues that, when combined, require a behaviour modification-oriented approach to prevention and harm reduction. Once this set of assumptions becomes the standard operating procedure for counter-radicalisation efforts, the work falls neatly into the communications domain built from the behaviour change premise and used widely in commercial advertising to understand choices, preferences, and habits. Unsurprisingly, layered underneath choice and preference are attitudes, presumptions, and behavioural intentions formed from perceptions gleaned from social exchanges and experiences [96]. From a purely theoretical perspective, it is fair to say that when looking at the role and purpose of CNs and, by default, communication efforts, the purpose of communication has to be seen as related to seeking some form of 're-socialising behavioural change' [3]. It is what this re-socialising process looks like in practice that is the main driver behind this project and its core research aim.

Some further discussion on this will take place in the main findings of the concluding chapter. The different theoretical approaches outlined in this chapter require some final bridging and synthesising to create the overarching conceptual framework of the project by arriving at a central hypothesis and supportive research enquiry routes. The need for behavioural change outcomes to be an essential element of P/CVE is acknowledged in what has been reviewed so far, but equally, what has been highlighted is that no tailored P/CVE-orientated model has been fully developed around the behaviour change and strategic communications-based intervention approach.

This anomaly has arguably had a significant impact on the role of strategic communications within P/CVE, as many assumptions about the role of strategic communications in this sphere are predicated on having a figure of authority or responsibility (government, community or tech sector) and an audience to reach. The term 'audience' implies a fundamental willingness to participate. If, on the other hand, personal agency has become the preferred 'pursuit', then telling someone something they are not interested in hearing is a failed strategy from the start. Extremists are active in various ways, and strategic communications practitioners and their employers are attempting to understand who to reach and with what content and platform. When one narrative emphasises personal agency, and the other emphasises tolerance and open-mindedness, an evident economy of scale emerges. Motivating someone is more likely to generate a return on investment than telling someone what they should or should not do. The concept of socialisation as it relates to the issue of radicalisation and, more precisely, to CN efforts and the objectives they may seek is undoubtedly an underdeveloped area of the field of terrorism studies. It is further removed from deradicalisation discourses in that little is known about the intended purpose of CN efforts before they enter into the ether of the online or offline space where they are supposed to take hold and have effect. These interlinked factors lead to a need to examine the extant literature and review the state of the P/CVE field as a sector in and of itself.

P/CVE Perspectives and Gaps in Understanding of the Sector

In further investigating how the preceding issues manifest within the P/CVE sphere, it is also necessary to acknowledge that people reject extremism, disengage, and desist from violence. The question highlighted in this review thus far is to what extent deradicalisation addresses and captures these types of behaviour change. Furthermore, has this been compared to self-disengagement or self-deradicalisation processes to arrive at a point in terrorism research where it adequately captures the relative contribution of self-realisation to preventative initiatives [83]. These points are critical to comprehending a values-based P/CVE policy and its effectiveness in countering the argument that these 'values' include military force and collateral damage. Thus, they are part of the problem rather than the solution. The simple Jihadi mantra of 'kill them as they kill you' appears to negate the allure of prevention and deradicalisation efforts in this context. While the mantra is based on authentic Quranic sources, terrorist groups influenced by Jihadi ideology disregard the narrative's contextual, historical, and theological implications. The majority of the world's Muslims have some basis for understanding their faith through some form of consensus-based paradigm and thus place such edicts in their 'proper' chronological order of real-world events [2].

However, as Edward Said famously stated, it must be acknowledged that much Western academic understanding, and thus its CT discourse, is implicitly framed in an orientalist view of the 'other' [107, 120]. In addition, the state bias that appears to permeate much of the research is obliviously ignoring critical push and pull factors (examples; foreign policy and the War on Terror legacy) to legitimise the foundations of post-modern liberal democracies that increasingly rely on neo-conservative idioms to pursue integration on the one hand and P/CVE on the other. The rise of malicious actor-based security, digital disruption, and disinformation threats demonstrates the slow and discernible disintegration of what constitutes a physical nation-state and, thus, the right to self-determination and protection versus the borderless, durable, and cross-sectional disorder that has characterised the twenty-first century [84].

There are very few alternative meta-narratives to orientalism-based contexts to make sense of specific comments that appear to imply that 'sympathy' for Jihadi causes correlates with some form of risk or threat. This chasm of divergent opinion within the P/CVE discourse is exemplified by Jenkins' view that the majority of

Western Muslims are hostile to Jihadi ideology. In contrast, Gartenstein-Ross argues that sympathy for the 'cause' should essentially be viewed as a threat to security [28]. By applying the equifinality principle to this issue, the broader question of how to manage homegrown threats becomes one of whether greater security implies greater peace or greater peace implies greater security. Adopting a P/CVE policy that encompasses both law enforcement and prevention measures essentially means that the candle is lit from both ends. Peace and security also appear to be relative terms in the context of an agenda that seeks absolute confirmation of its known and emerging adversaries [66].

With competing perspectives on P/CVE and a lack of robust conceptual clarity on radicalisation and deradicalisation issues, one can only conclude that P/CVE efforts in the counter-Daesh context are still in their infancy and will require additional time and, more importantly, further data to dispel both the negativity that surrounds them and the contextual amnesia that pervades much of its current discourse. Mellen argues that neither Jenkins' nor Gartenstein-Ross' perspectives can present the 'accurate' picture that both have sought to frame differently [86]. What is clear is that the nature of religiosity in contemporary society is underdeveloped within this discourse; thus, rather than painting a picture of a quasi-orientalist P/CVE understanding pervading the discourse, it is more effective to examine the role of religion in several Western societies in detail. This brings us to the study's primary focus, namely the role communication has played in P/CVE thus far, how extremists use communication to disseminate ideology via messaging, and the extent to which alternative and CNs have been or have not been fully understood.

Application of Concepts Within P/CVE Efforts

The P/CVE sector has been built on the presumption of being able to create intervention-based approaches to prevent radicalisation. From a sociological perspective, it is individuals who 'decide' whether to become extremists or not, and this decision is frequently made through subconscious processes and choices as well as intentional sets of methods and choices. Education, social mobility, social inclusion, and support networks, among other factors, do not guarantee that an individual will not become an extremist. In other words, resilience does not guarantee that an individual will not adopt extremist ideas and neither do initial phases of socialisation. Far too many variables are involved for resilience on its own to be an antidote to radicalisation. All evidence indicates that people are either persuaded or compelled to adopt extremist ideas.

This is where the concept of socialisation becomes helpful in pining down radicalisation into some form of theoretical paradigm. How resilient people are or are not has no bearing on their choice or preference trajectory in isolation; it only affects their level of engagement with the ideology or set of narratives that eventually convinces them to become extremists through some form of socialisation. At the individual level, resilience affects narratives and notions of selfhood; at the social level, it influences narratives of community and collectivism. Extremism provides both, albeit in distinct forms, so is it not counter-intuitive for P/CVE efforts to continue emphasising resilience building as a primary mode of delivery?

In essence, if someone gains a greater sense of self-awareness and resilience as a result of extremist ideology, will offering them additional training in resilience or critical thinking deter them? Suppose the prevailing social conditions contributed to this person's problem-set. How are these same social conditions or inputs based on generic concepts of resilience and critical thinking going to magically 'fix' them? This is where counter-radicalisation fails and serves as a precursor to why deradicalisation is also seen more as a loose concept when analysed and assessed strategically. As a result, one could argue that deradicalisation is a more contentious concept than radicalisation. At the most fundamental level, deradicalisation can be crudely defined in Neumann's language as the process by which a bomb does not detonate [94]. Still, as is self-evident, this is far too simplistic and misleading for various reasons. This crude statement does, however, emphasise more briefly the importance of conceptual clarity for policy and operational discourses in the P/CVE world to be in some form of synergy. This appears to be a significant issue for the language lexicon that still pervades the P/CVE world, as evidenced by a literature review of the post-9/11 discourse development.

To understand why some people choose to abandon extremism, too much academic emphasis has been placed on psychological-emotional processes at the expense of communication and persuasion in terms of meanings and narratives. Due to a lack of access to raw data, longitudinal research, and an odd aversion to ignoring the central role of ideology in both processes, the integrity of the available evidence is called into question. As Robert Pape once suggested, there is a strategic logic to why people leap from ideology to violence; an element of this logic is based on the power of persuasion [97]. Strategic communications have been grossly undervalued in the broader discourse. If, for instance, the RAM is used to analyse the culmination of knowledge, attitudes, and behaviours in conjunction with the widely accepted socialisation model, this provides a conceptual framework capable of incorporating personal, public, and ideological manifestations of 'change' dynamics.

Additionally, this approach incorporates sufficient space within its conceptual framework to consider the role of the internet, mass communication, the rise of individualism and extremism, as well as the theoretical requirements for what constitutes counter-messaging, CNs, and counter-radicalisation in terms of communication through policy and practice. Several important issues have been highlighted in this review. A concluding discussion that combines the main gaps, problem areas, and threads is helpful for this study in building an appropriate theory framework for the next chapter.

According to the findings of this review, the impact of the 'war on terror' and 'kinetic force' models of military intervention within a CT context appear to be underdeveloped areas of current radicalisation and deradicalisation research. Another issue raised by this research effort is how, even when the terms radicalisation and deradicalisation are given a relative meaning, it is determined by comparing them to the terms 'moderate' or 'mainstream'. This establishes the ground for another anomaly in contemporary policy-based understanding, which Herbert best illustrates when he states:

> "in reality, a thousand flowers are blooming in the realm of Islamic activism: approaching such diversity with a single us-or-against-us dichotomy in mind is a hopeless, futile task" [56].

The gradual rise of extremist factions in the United States, culminating in the Washington attacks in 2022, demonstrates that what was once considered mainstream is now barely mainstream unless conspiratorial extremism and othering have become mainstream. While some academic perspectives can offer and present these nuances in a more balanced manner, these perspectives appear to be few compared to the vast array of conceptual models examined in this study. According to Githens-Mazer, radicalism must be contrasted with apathetic tendencies as opposed to activism [47]. This is a helpful perspective, but it makes some assumptions about the 'Muslim mindset'. It echoes many of the criticisms it appears to be attempting to refute, namely that concepts of radicalisation are too broad and based on simplistic typologies that, in reality, are far more complex and nuanced than any single concept can capture.

By contrasting radicalism and apathy, does this not imply that 'apathetic' Muslims can also become vulnerable to extremism via the more 'accepted' mode of activism? Suppose Islamic 'activism' is not viewed as a threat. In that case, non-violent Islamist groups are the erroneous target of the 'integration'-based P/CVE agenda, owing to their preference for non-violent methods of social change over violent ones. In this view, apathy is not a state of mind in comparison to 'Islamism', as the former requires only specific push and pull factors to progress through its stages (grievance, motivation, and action). Apathy is not an ideology, whereas the constant drive to subdivide Islam into 'schisms' and 'isms' indicates that either the true nature of the threat posed by Jihadi ideology is still unknown or that terminology is inadvertently confounding P/CVE efforts.

There can be little doubt that Jihadi ideology possesses the capacity to operate within the broad and fundamental community structures of Muslim diasporas, as it does so through the use of critical theological lexicons. P/CVE efforts seek to mitigate this 'home advantage' by framing risk and vulnerability to extremism/terrorism from within the same 'sample' (general Muslim populations), but from an external position of influence and by employing terminology that erodes the group's 'outsider' label. Without theological clarity, doctrinal consistency, cultural synergy, and 'trust', P/CVE language and thus communication appear to be inadvertently increasing vulnerability to extremism rather than countering it. As previously stated in this study, there are numerous manifestations of Jihadi ideology's 'self' at the moment. Given the relationship between online or self-radicalisation and established conceptual frameworks, this point can be applied to various velocities of radicalisation processes and multiple motivations for their journeys [6].

Although a helpful distinction, Kuhle and Lindekilde's assertion that 'true' radicalisation must be distinguished from 'expressions of youthful frustration, revolt, and solidarity with Muslim world populations' also casts some doubt on the current UK context of a values-based preventative CT policy [73, 115]. One could argue that a values-based approach must also acknowledge the relative nature of what 'values' mean and that once concepts such as radicalisation are added, achieving absolute conceptual clarity and security becomes difficult. As Mandel et al. points out, the term 'radical' is distinct from the term 'terrorist ', that 'radical is generally defined in terms of the extremist', and the majority of definitions regard the 'radical' as a threat [81]. This element of 'attributions' in defining radicalisation is further confused by acknowledging that it does not always result in violence because, simply put, "radicalisation cannot be a sufficient cause of terrorism because the vast majority of radicals are not terrorists" [81].

In Reich's 'The Origins of Terror', Crenshaw brilliantly captures this growing uncertainty by tracing the evolution of the term from one emphasising the intrinsically 'fundamental' to one emphasising the 'extreme' [105]. There is also the issue of the concept's continued evaluative understanding and whether such efforts and concepts can firmly identify the motivations for terrorism by framing radicalisation in value-laden subjectivism [81]. Perhaps most critically, from a conceptual standpoint, radicalisation discourse is currently incapable of explaining 'how that process unfolds or why it begins' [81]. Indeed, Mandel et al. argue that there is still a need to distinguish between the structure of extremist thinking and its contents when attempting to express radicalisation processes from a cognitive perspective [81]. Contemporary discourse appears far from resolving these issues, with much of the academic and practitioner focus on establishing their respective 'proof of concepts'. Hillyard's research on the impact of UK CT policy on the Provisional Irish Republican Army threat over the preceding decades indicates that despite grave 'perceived grievances', most of the Irish community in the UK and Ireland 'managed' these grievances without becoming radicalised or 'extreme'. Human capacity for pain thresholds varies enormously, and thus, the way grievances affect each individual cannot be attributed to a single causal factor, regardless of the definition of radicalisation or terrorism.

Mandel et al. argue that the connotative meaning of the term radicalisation should ultimately be understood as an increase in extremism rather than as a process. Thus, even from a 'structuralist semiotic' perspective, the term is currently used to describe an incremental cognitive rise in extremism, not necessarily the mechanism by which this occurs. While radicalisation concepts refer to a process, the term's connotation is based on the emergence of extremist tendencies, manifested in some cases through violence. As Zaidi points out, this has real-world ramifications for the term 'radicalisation', as the term is defined in the real world only by its association with extremism in the context in which it is used;

> "Nobody wants to be labelled an extremist, regardless of how radical their worldview, and thus the term lacks context through ownership" [122].

This may be why Sedgwick's point about the three lenses through which the term is used limits the term's operational efficacy when issues of religiosity and conservatism in religious practice are introduced. In this context, Zaidi argues that religious zeal and militant narratives are traditionally associated with some form of 'retrogressive barbarism' and a general lack or absence of a 'civilised' or 'modern' worldview [122]. However, with such a framework in place, there is a risk that the term will be interpreted as reinforcing equally regressive orientalist or colonial binaries when applied to issues that require more nuanced leverage to be effective [122]. Thus, while radicalisation discourse may carry meaning within the CT context, it may fall short of doing so within the 'real world' dialectic of terrorism it seeks to comprehend.

If an increase in religiousness (not extremism) is viewed internally as a beneficial process but externally as a result of radicalisation, it is argued that the typology lacks an adequate frame of reference for describing what moderate religiousness should look like, except within the already skewed view of acceptance of 'Western values' and so forth. However, suppose the recipient follows a non-violent form of religious understanding instead of democratic principles and integration. How can current radicalisation discourse frame its stance without labelling the individual as an extremist? As previously stated, if the radical is viewed as a threat and the term is used in conjunction with extremism, then a non-violent, religiously conservative individual remains a potential risk, according to the current policy understanding.

This study previously stated that a simple lexicon of language is frequently used to 'trigger' radicalisation processes. Identity and belonging are pervasive themes in Jihadi-influenced literature and narrative. However, as King and Taylor explain, the outcome of any such process in this context of identity crisis is determined by the 'person and the situation' [70]. Can this be taken to mean that someone who is religiously observant is just as likely, if not more likely, to be radicalised as someone who is newly introduced to religious praxis through radicalisation? If this is what it means, the discourse fails to capture the essence of the distinction between increased religiosity and radicalisation, let alone distinguish between these nuances. There are profound implications for P/CVE efforts and the role of strategic communications on this point alone.

Numerous radicalisation models presented in this study's literature review appear to contain conceptual and assumptive issues. Mellen, for example, distinguishes two strikingly similar radicalisation models [103, 111]. Silber and Bhatt, she argues, include Salafism in the 'necessary' components of the radicalisation stages preceding acceptance of violent Jihad. In contrast, Precht is more ambiguous, citing Islam as a means to an end. The fact that two similar models arrive at the same conclusions about radicalisation processes but offer disparate explanations for the prevalent factors demonstrates the pervasive conceptual confusion in P/CVE and CT understanding [86]. This is even more evident in Precht's assertion that those seeking a more pronounced 'Muslim identity' are inevitably drawn to 'radical Islam' and thus radicalised [103]. This claim is perhaps best understood because Precht's study drew on publicly available data rather than primary empirical research sources. Making such an assertion without primary data gathered from a wide demographic field opens it up to criticism of reliability and suitability for drawing both general and specific conclusions.

As previously stated, this point continues to be the Achilles heel of much terrorism studies research, namely the lack of access to high-quality primary sources for empirical analysis [86]. Mellen asserts, 'Precht fails to investigate which combinations are most effective at inciting radicalisation, or whether any are necessary or sufficient; additional research in this area is warranted' [86]. Jenkins argues that for some the 'action-oriented' allure of extremist ideology is more compelling than 'finding a spiritual connection to Islam' [66]. This perspective does appear to distance the religious observer from the potential terrorist, as it follows a path that can be correlated along predisposed and situational paradigms without placing an excessive emphasis on what radicalisation means to it. While most definitions of radicalisation include faith interpretation as a critical component of any process, the fundamental question of what constitutes a specific interpretation of faith and why it may still affect people differently remains unexplored mainly at the moment [86]. The absence of direct access to empirical data and 'radicalised' subjects is a criticism levelled at even the most innovative exposes of Jihadi ideology, such as Marc Sageman's work, whose findings have been questioned for this very reason [28].

When combined with the emerging and ever-evolving threat of online radicalisation, one could argue that the process can be stretched out over years on the one hand and compressed into hours or days on the other. To this extent, the shape and structure of Daesh (post-territorial defeat) are irrelevant when the discourse used to unearth said capability is riddled with conceptual amnesia. Terrorism research has seen a significant increase following 9/11. The question, however, must be asked: how much of this research is still based on pre-9/11 terrorism studies paradigms to explain even the internet-based threat from a globalist terrorist network that is vastly different from a millenarian cult or gang in terms of scale and structure Current discourse does not appear to be able to articulate the significance of traditional one-to-one recruitment in comparison to virtual recruitment or how both formal and informal modes of indoctrination contribute to increased or decreased vulnerability to violence. The fact that much of the discourse continues to use the term Salafi or Salafism demonstrates that radicalisation is not only a problematic concept but also the terminology associated with it. For example, Githens-Mazer refers to violent Takfiri Jihadis, the NYPD to Salafism, and Spencer, Precht, and others to Islam. This is a concerning feature and one of the study's key findings on the term radicalisation, as it reveals a great deal about the role, or lack thereof, that effective communication has played or should have played.

Summary of Socialisation, Communicative Behaviour Change Models and P/CVE Sector

Whether P/CVE CN strategic communications are counter-intuitive or fit for purpose stems from a seemingly underdeveloped area of work in terrorism studies. This underdevelopment is based on the lack of any core theoretical contribution to the issue of the use, place, purpose and assumed effectiveness of strategic communications practices as manifested in online counter-radicalisation efforts. Issues relating to the sensitive nature of this work and the constraints placed on evaluation and analysis through lack of access to critical data are acceptable limitations in some sense, but what is surprising is the absence of conceptual framing and critical enquiry into the logic and pedagogy of such communication efforts to both improve and advance the field both academically and in the area of operations through continuous review and good practice. With this in mind, the following theoretical summary of key concepts discussed in this chapter is now presented as an overarching discussion that could assist with the formation of a suitable theory framework for this research project to base its key hypothetical assumptions on in Chap. 2.

The concept of political socialisation offers a robust and developmental foundation with which to understand both the use of socialisation techniques by extremists to engage with audiences as well as how nation-state actors and their stakeholders, partners and allies seek to re-embed the dominant norms and values of society upon its 'members' [31]. The use of the term 'citizens' is quickly replaced by the term 'members', and in doing so, the central premise becomes one of citizenship of the state, meaning membership of the society, and this requires the individual citizen (member) to abide by and respect the political, legal, and social structures in place. Failure to do so results in dysfunction at the social level and potential disassociation with said norms and values at the cognitive level through adopting attitudes and perceptions that cause certain undesirable behaviours [36]. This results in issues of self-identity and social identity management through stigma and spoiled identity, further exacerbating the sense of 'othering' and social isolation [43, 49]. At this point in the post-modern era, the level of vulnerability created becomes a key weapon for those seeking support and followers for different ideological and social change narratives, such as extremist ideas [52].

The issue with the political socialisation approach is that it appears to favour those contexts where violence is not an inherent part of the modus operandi of the network, organisation, and movement in question. Violent ends may emerge, but not directly from ideological leanings towards its use at the outset. Political socialisation implies a form of social mobilisation that may well stay within the socio-political confines of a particular worldview rather than seek violent social change. This means that for understanding trajectories and pathways that lead to extremism/violent extremism, focusing on transmitting, adopting, and adapting norms and values through the parent theory and socialisation may be more effective. This may provide a more relevant perspective to situate contemporary issues linked to online extremism and prevalent narratives. The rate and scope of change in how communication, identity and participation are now considered require an understanding of the global digital era that social media, tech and connectivity have created.

The rapid technological changes that have ushered in a new global age based primarily on immensely versatile and borderless forms of information exchange and commerce have resulted in second-order effects on the legitimacy and authority perceptions of the classic Westphalian nation-state model and the liberal democratic ideas that have gradually underpinned this form of governance [84]. Greater access to information and audiences, alongside the proliferation of war and instability since the break-up of the Age of Empire, combined with immense technological power regarding modern communication systems, has brought a stark contrast to the developed world. Recent manifestations of this development, where kinetic force, lifestyle, life chances, and grievances are now aired and shared in a global theatre of public spectacle, have allowed binary narratives (them and us) and othering to take on mainstream manifestations through events like the 2016 US elections and EU Brexit vote in the UK. Such events harnessed and exploited the use of instrumentarianism made available by increasingly powerful and 'stateless' or 'borderless' tech corporations who have become both provider/protagonist and influencer for the global 'service user' [123]. Traditional forms of socialisation have been replaced by artificial intelligence-led decision-making and communication methods that have coincided with the rise of 'influencer' culture. The nation-state now grapples with social media influencers to maintain both visibility and elements of legitimacy. At the same time, extremist groups and malign actors actively distort narratives to suit their own network's strategic needs. Social action has now taken a fragmented persona, whereby different social/political interest groups jockey for attention and legitimacy at the expense of each other and often forsaking facts for 'likes' or 'shares' or other vanity metrics [108].

Terrorist networks are now as dispersed globally as the millions of migrants seeking new lives in the developed world. They have been beamed into their smartphones and social media accounts through this same architecture of global influencer networks, tales of 'normal' people telling 'stories' [79]. These narratives are morphing daily, and alongside the surge in tech and its use in everyday life, the nature of the terrorist threat has changed. The live-streaming of the Christchurch attack in 2019 serves as a warning call to both the tech sector and governments of the world that both the modus operandi of terrorists have changed and taken on more of the facets of the virtual world society now inhabits. The messages of extremism have become increasingly sophisticated, whilst its language remains chillingly simple [79]. This raises some fundamental questions around the gamification of terrorism, the mainstreaming of extremism into everyday social narratives, and the effectiveness of efforts to counter-extremist narratives through targeted and strategically orientated communications.

In an era of 'post-truth', where alternative facts are considered viable tools for political and social change, the central question is: are CNs counter-intuitive simply because they cannot control behaviour or shift attitudes with the current approach being used? This approach to counter extremism using communications is loosely viewed as somewhat clandestine/unattributable and assumption-based, with aims on the ability of a short-medium-term communications campaign to impact individual mindsets significantly. These cognitive and behavioural factors exist with enormous nuance and situational variance as standard. The ongoing debate on whether ideology or environment is the cause of extremism magnifies the confusion. It obfuscates already confusing communications aims regarding the real purpose of these efforts or what they are trying to achieve [37]. This study aims to shed light on these matters to move the discourse and understanding of matters forward more effectively. A unique methodological approach is needed to shed new light on the P/CVE CN paradigm for this aim.

Considering this theoretical set of contexts, it is clear that socialisation theory and communication studies become vast, expansive discourses that would ultimately require more refinement and structuration to benefit P/CVE as a whole. When situated alongside extremist ideology, narrative, and potency, the immense subjects of social processes and political and cultural narratives create a unique set of needs for the P/CVE sector. These can be best summarised in a short outline of what the overarching model can offer and how this might be applied through continued research and development to refine parameters and contexts.

Conclusion

As seen by this chapter, diverse approaches towards radicalisation and deradicalisation and their heterogeneity must be fully understood to deploy P/CVE tools effectively. As such, socialisation theory, behaviour change models and communication studies present themselves as precursors to these discourses and possible alternative models. They can facilitate this heterogeneous and highly nuanced field of understanding.

This chapter has reviewed the extant literature around the theoretical frameworks and methodologies that make up this enquiry. Radicalisation and deradicalisation discourses are essential but need to be understood and researched further alongside a synthesis of socialisation and social processes and not as separate entities. Socialisation theory can appear very broad, but this offers a chance to change the lack of consensus in radicalisation and deradicalisation discourse to embrace nuance and heterogeneity more openly and for the benefit of the P/CVE sector overall. In doing so, the nexus of extremism and its ever-changing manifestations can be identified with both a broad social change lens and drilling down into specific remnants of online, offline and hybrid variants of extremist ideology.

Such an approach would allow P/CVE to be both reactive and proactive in addressing prevention needs in a more future-proof manner through placing social processes and individual pathways to extremism as accepted features and not outlier problems with only conjecture or 'hit and miss/quick-win' counter-efforts on offer. These elements can then merge to form a more behaviour change-centric focus for

P/CVE efforts that encompasses the personal, social, ideological, and technological aspects on their own merits and not as an all-encompassing endeavour.

This review emphasised the importance of concurrently examining socialisation processes, strategic and generic communication models of behaviour change, and P/CVE assumptions to develop a foundational understanding of the future needs of this work. It has done so by uncovering flaws with the discourses on radicalisation, which leads to a set of critical issues emerging for the research component of this study to focus on moving forward. The next chapter explains the theoretical frameworks behind this analysis and how these can be adapted to suit P/CVE needs more holistically. The operationalisation of socialisation and communication studies will be examined as possible theoretical frameworks to understand alternative approaches to existing practice within P/CVE. The objective of this exercise is to create a more holistic model for P/CVE communications seeking behaviour change-based outcomes by outlining possible benefits and any limitations.

References

1. Abu-Nasr, D, & Keath, L. (2009). *HuffPost—Breaking News, U.S. and World News*. [online] HuffPost. Available at: http://www.huffingtonpost.com/2009/11/19/increasing-number-of-web_n_364635.html [Accessed 19 Mar. 2024].
2. Aftab Ahmad Malik. (2001). *The Broken Chain*. Amal Press.
3. Ajzen, I. (1991). The theory of planned behavior. *Organizational Behavior and Human Decision Processes, 50*(2), 179–211. https://doi.org/10.1016/0749-5978(91)90020-T
4. Anderson, R. (2016). The Rashomon Effect and Communication. *Canadian Journal of Communication*, 41(2). https://doi.org/10.22230/cjc.2016v41n2a3068.
5. Archetti, C. (2013). *Understanding terrorism in the age of global media: a communication approach*. Palgrave Macmillan.
6. Ashour, O. (2010). Online de-radicalization? Countering violent extremist narratives: Message, messenger and media strategy. *Perspectives on Terrorism*, [online] 4(6), pp.15–19. Available at: https://www.jstor.org/stable/26298491.
7. Baker-Beall, C., Heath-Kelly, C., & Jarvis, L. (2016). *Counter-radicalisation: Critical perspectives*. Routledge.
8. Bauman, Z. (2000). *Liquid Modernity*. Polity Press.
9. Bauman, Z. (2006). *Liquid fear*. Polity Press.
10. Baumeister, R. F., Masicampo, E. J., & DeWall, C. N. (2009). Prosocial benefits of feeling free: Disbelief in free will increases aggression and reduces helpfulness. *Personality and Social Psychology Bulletin, 35*(2), 260–268. https://doi.org/10.1177/0146167208327217
11. Beck, U. (2009). *World at risk*. Polity Press.
12. Berlo (n.d.). *Figure 1. Berlo's (1960) Source-Message-Channel-Receiver Model of...* [online] ResearchGate. Available at: https://www.researchgate.net/figure/Berlos-1960-Source-Message-Channel-Receiver-Model-of-Communication-Developed-from_fig1_331435468.
13. Best, S. (2003). *A beginner's guide to social theory*. Sage.
14. Borum, R. (2003a). *Understanding the terrorist mind-set*. [online] Available at: https://www.ojp.gov/pdffiles1/nij/grants/201462.pdf.
15. Borum, R. (2003b). Understanding the Terrorist Mindset. *FBI Law Enforcement Bulletin*, [online] 72(7–10). Available at: https://works.bepress.com/randy_borum/7/ [Accessed 19 Mar. 2024].
16. Borum, R. (2011). Radicalization into violent extremism i: A review of social science theories. *Journal of Strategic Security, 4*(4), 7–36. https://doi.org/10.5038/1944-0472.4.4.1

17. Bourdieu, P. (2023). *Habitus and Field: General Sociology, Vol 2 (1982–1983)*. Polity Press.
18. Braddock, K. (2020). *Weaponized Words: The strategic role of persuasion in violent radicalization and counter-radicalization*. Cambridge University Press.
19. Brim, O. G., & Wheeler, S. (1976). *Socialization after childhood*. John Wiley and Sons.
20. Brown, D., &Malik, S. (2007). *7/7: looking back*. [online] www.prospectmagazine.co.uk. Available at: https://www.prospectmagazine.co.uk/politics/uk/54282/77-looking-back [Accessed 22 Mar. 2024].
21. Burgoon, M., Hunsaker, F. G., & Dawson, E. J. (1994). *Human communication*. SAGE Publications.
22. Burke, J. (2007). *Al-Qaeda: the true story of radical Islam*. Penguin.
23. Carey, J. W. (1992). *Communication as culture: Essays on media and society*. Routledge.
24. Carey, J.W. (2008). *Communication as Culture, Revised Edition*. Routledge.
25. Gefwert, C. (2017). *Wittgenstein on thought*. Routledge.
26. Cox, M. H. (1959). Maxwell Anderson and composition and communication. *College Composition and Communication, 10*(4), 239. https://doi.org/10.2307/355099
27. Danziger, K. (1973). *Socialization*. Penguin Books.
28. Gartenstein-Ross, D., Goodman, J. D., & Grossman, L. (2009). *Terrorism in the west 2008: A guide to terrorism events and landmark cases*. Fdd Press.
29. Dawson, L.L. (2019). Taking terrorist accounts of their motivations seriously: An exploration of the Hermeneutics of suspicion. *Perspectives on Terrorism, 13*(5), 65-80.
30. Dewey, J. (1997). *Experience And Education*. Kappa Delta.
31. Durkheim, E. (1956). *Education and sociology*. Free Press.
32. Dutta-Bergman, M. J. (2005). Theory and practice in health communication campaigns: A critical interrogation. *Health Communication, 18*(2), 103–122. https://doi.org/10.1207/s15327027hc1802_1
33. Ephraim, P. (2015). *Book Review of Understanding Terrorism in the Age of Global Media*. [online] International Journal of Communication. Available at: https://ijoc.org/index.php/ijoc/article/viewFile/3848/1334 [Accessed 12 Jan. 2022].
34. Falkheimer, J. and Heide, M. (2011). *Participatory strategic communication: From one- and two-way communication to participatory communication through social media: ICA 2011 – Annual Conference of the International Communication Association, Pre-conference: Strategic communication – A concept at the center of applied communications*. [online] Lund University. Available at: https://portal.research.lu.se/en/publications/participatory-strategic-communication-from-one-and-two-way-commun [Accessed 22 Mar. 2023].
35. Fishbein, M., & Ajzen, I. (1975). *Belief, attitude, intention, and behavior: An introduction to theory and research*. Addison-Wesley Pub. Co.
36. Foucault, M. (1961). *Madness and civilization: A history of insanity in the age of reason*. Routledge.
37. Francis, M. (2014). *How do we prevent radicalisation? By Matthew Francis*. [online] Theos Think Tank. Available at: https://www.theosthinktank.co.uk/comment/2014/09/26/how-do-we-prevent-radicalisation-by-matthew-francis [Accessed 20 Mar. 2022].
38. Fukuyama, F. (2012). *The end of history and the last man*. Penguin.
39. Fukuyama, F. (2018). *Identity: Contemporary identity politics and the struggle for recognition*. Profile Books.
40. Fukuyama, F. (2020). *The Future of History*. [online] www.foreignaffairs.com. Available at: https://www.foreignaffairs.com/articles/2012-01-01/future-history [Accessed 27 Mar. 2022].
41. Furedi, F. (2007). *Invitation to terror*. Continuum.
42. Furedi, F. (2008). Fear and security: A vulnerability-led policy response. *Social Policy & Administration, 42*(6), 645–661. https://doi.org/10.1111/j.1467-9515.2008.00629.x
43. Garfinkel, H. (1967). *Studies in ethnomethodology*. Prentice Hall.
44. George Herbert Mead (2015). *Mind, Self, and Society*. University of Chicago Press.
45. Giddens, A. (1984). *The constitution of society: Outline of the theory of structuration*. Polity Press.

46. GIFCT (2021). *Content-Sharing Algorithms, Processes, and Positive Interventions Working Group Part 2: Positive Interventions*. [online] Available at: https://gifct.org/wp-content/uploads/2021/07/GIFCT-CAPI2-2021.pdf [Accessed 18 Dec. 2022].
47. Githens-Mazer, J., & Lambert, R. (2010). Why conventional wisdom on radicalization fails: The persistence of a failed discourse. *International Affairs, 86*(4), 889–901. https://doi.org/10.1111/j.1468-2346.2010.00918.x
48. Glazzard, A. (2017). *Losing the Plot: Narrative, Counter-Narrative and Violent Extremism*. [online] International Centre for Counter-Terrorism—ICCT. Available at: https://www.icct.nl/publication/losing-plot-narrative-counter-narrative-and-violent-extremism.
49. Goffman, E. (1963). *Stigma: Notes on the management of spoiled identity*. Penguin Books.
50. Goli, M. and Rezaei, S. (2010). *House of War: Islamic Radicalization in Denmark*. [online] Aarhus University. Available at: https://www.ps.au.dk/fileadmin/site_files/filer_statskundskab/subsites/cir/Rapport_2FINAL.pdf [Accessed 9 Apr. 2022].
51. Hallahan, K., Holtzhausen, D., van Ruler, B., Verčič, D., & Sriramesh, K. (2017). Defining strategic communication. *International Journal of Strategic Communication, 1*(1), 3–35.
52. Hassan, A. (2012). Shifting paradigms. *Arches Quarterly*, 5(9).
53. Heath-Kelly, C., & Strausz, E. (2018). The banality of counterterrorism 'after, after 9/11'? Perspectives on the Prevent duty from the UK health care sector. *Critical Studies on Terrorism, 12*(1), 89–109. https://doi.org/10.1080/17539153.2018.1494123
54. Hedges, P. (2017). Radicalisation: Examining a concept, its use and abuse. *Counter Terrorist Trends and Analyses*, [online] 9(10), pp.12–18. Available at: https://www.jstor.org/stable/26351560?seq=1. [Accessed 21 Nov. 2021].
55. Hegghammer, T. (2010). The rise of muslim foreign fighters: Islam and the globalization of Jihad. *International Security*, [online] 35(3), pp.53–94. Available at: https://www.jstor.org/stable/40981252.
56. Herbert, M. (2009). The plasticity of the Islamic activist: Notes from the counterterrorism literature. *Studies in Conflict & Terrorism, 32*(5), 389–405. https://doi.org/10.1080/10576100902827972
57. Hobbis, I. C. A., & Sutton, S. (2005). Are techniques used in cognitive behaviour therapy applicable to behaviour change interventions based on the theory of planned behaviour? *Journal of Health Psychology, 10*(1), 7–18. https://doi.org/10.1177/1359105305048549
58. Hoffman, B. (2017). *Inside terrorism* (3rd ed.). Columbia University Press.
59. Horgan, J. (2008a). Deradicalization or disengagement?: A process in need of clarity and a counterterrorism initiative in need of evaluation. *Perspectives on Terrorism*, [online] 2(4), pp.3–8. Available at: https://www.jstor.org/stable/26298340.
60. Horgan, J. (2008). From profiles to pathways and roots to routes: perspectives from psychology on radicalization into terrorism. *The ANNALS of the American Academy of Political and Social Science, 618*(1), 80–94. https://doi.org/10.1177/0002716208317539
61. Horgan, J. (2016). A call to arms: The need for more psychological research on terrorism. *Social Psychological Review, 18*(1). Spring.
62. Horgan, J., Altier, M. B., Shortland, N., & Taylor, M. (2016). Walking away: The disengagement and de-radicalization of a violent right-wing extremist. *Behavioral Sciences of Terrorism and Political Aggression, 9*(2), 63–77. https://doi.org/10.1080/19434472.2016.1156722
63. Horgan, J. and Braddock, K. (2012). *Terrorism studies: a reader*. Routledge.
64. Institute for Strategic Dialogue (2016). *The Impact of Counter-Narratives*. [online] ISD. Available at: https://www.isdglobal.org/isd-publications/the-impact-of-counter-narratives/#:~:text=ISD [Accessed 19 Nov. 2022].
65. Jackson, R. (2016). *Routledge handbook of critical terrorism studies*. Routledge.
66. Jenkins, B.M. (2010). *Would-Be Warriors: Incidents of Jihadist Terrorist Radicalization in the United States Since September 11, 2001*. [online] www.rand.org. Available at: https://www.rand.org/pubs/occasional_papers/OP292.html [Accessed 18 Nov. 2022].
67. Khoshnood, A. (2017). The correlation between mental disorders and terrorism is weak. *BJPsych Bulletin, 41*(1), 56–56. https://doi.org/10.1192/pb.41.1.56

68. Kaiser, F. G., Schultz, P. W., & Scheuthle, H. (2007). The theory of planned behavior without compatibility? Beyond method bias and past trivial associations. *Journal of Applied Social Psychology, 37*(7), 1522–1544. https://doi.org/10.1111/j.1559-1816.2007.00225.x
69. Kaunert, C., & Leonard, S. (2013). *European security, terrorism and intelligence Tackling new security challenges in Europe*. Palgrave Macmillan.
70. King, M., & Taylor, D. M. (2011). The radicalization of homegrown Jihadists: A review of theoretical models and social psychological evidence. *Terrorism and Political Violence, 23*(4), 602–622. https://doi.org/10.1080/09546553.2011.587064
71. Kippax, S. (2017). A journey to HIV prevention research: From social psychology to social health via multidisciplinarity. *Journal of Health Psychology, 23*(3), 442–456. https://doi.org/10.1177/1359105317707529
72. Kohn, M. L. (1973). Social class and schizophrenia: A critical review and a reformulation. *Schizophrenia Bulletin, 1*(7), 60–79. https://doi.org/10.1093/schbul/1.7.60
73. Kuhle, L. and Lindekilde, L. (2010). *Radicalization Among Young Muslims in Aarhus*. [online] Available at: http://tinyurl.com.bronby2 [Accessed 5 Jun. 2021].
74. Lee, T.T. (2021). Social class, intensive parenting norms and parental values for children. *Current Sociology*, p.001139212110485. https://doi.org/10.1177/00113921211048531.
75. Leviton, L. C., & Trujillo, M. D. (2016). Interaction of theory and practice to assess external validity. *Evaluation Review, 41*(5), 436–471. https://doi.org/10.1177/0193841x15625289
76. Lofland, J., & Skonovd, N. (1981). Conversion motifs. *Journal for the Scientific Study of Religion, 20*(4), 373. https://doi.org/10.2307/1386185
77. Wittgenstein, L., & Wright, V. (2006). *Culture and value*. Chicago Univ. Press.
78. Macheke, C., & Campbell, C. (1998). Perceptions of HIV/AIDS on a Johannesburg Gold Mine. *South African Journal of Psychology, 28*(3), 146–153. https://doi.org/10.1177/008124639802800304
79. Maan, A. (2018). *Narrative Warfare*. Createspace Independent Publishing Platform.
80. Ranstorp, M. (2010). *Understanding violent radicalisation*. Routledge.
81. Mandel, D. (2009). *Radicalization: What does it mean?* [online] Available at: https://citeseerx.ist.psu.edu/document?repid=rep1&type=pdf&doi=12e7112898a6e104de030c0279ec25b2303190a9.
82. McCauley, C., & Moskalenko, S. (2008). Mechanisms of political radicalization: Pathways toward terrorism. *Terrorism and Political Violence*, [online] 20(3), pp.415–433. https://doi.org/10.1080/09546550802073367.
83. Mcdonald, L. Z. (2011). Securing identities, resisting terror: Muslim youth work in the UK and its implications for security. *Religion, State and Society, 39*(2–3), 177–189. https://doi.org/10.1080/09637494.2011.584712
84. Mcfate, S. (2019). *The new rules of war: Victory in the age of durable disorder*. William Morrow, An Imprint Of Harpercollins Publishers.
85. McIntyre, L. (2018). *Post-Truth*. MIT Press.
86. Mellen, C. (2012). When Harry Met Salafi: Literature review of homegrown Jihadi terrorism. *Journal of Applied Security Research, 7*(2), 239–252. https://doi.org/10.1080/19361610.2012.656257
87. Merton, R.K. (1938). Social structure and anomie. *American Sociological Review*, [online] 3(5), pp.672–682. https://doi.org/10.2307/2084686.
88. Mitchuk, O, Penchuk, I., Podluzhna, N., Malovicko, O,. Shirobokova, O,. & Tergub A. (2021), Changes in social communication as a tool of social work under the influence of digitalization. *Studies of Applied Economics, 39(3)*. https://doi.org/10.25115/eea.v39i3.4717
89. Miller, G. A. (1951). *Language and communication*. McGraw-Hill. https://doi.org/10.1037/11135-000
90. Moghaddam, F.M. (2005). The staircase to terrorism: A psychological exploration. *American Psychologist*, [online] 60(2), pp.161–169. Available at: http://fathalimoghaddam.com/wp-content/uploads/2013/10/1256627851.pdf.
91. Moghaddam, F. M., & Marsella, A. J. (2004). *Understanding terrorism: Psychosocial roots, consequences, and interventions*. American Psychological Association.

92. Morrison, J., & CREST (2017). *A systematic review of post-2017 research on disengagement and deradicalisation.* [online] crestresearch.ac.uk. Available at: https://crestresearch.ac.uk/resources/a-systematic-review-of-post-2017-research-on-disengagement-and-deradicalisation/.
93. Morrison, J.F. (2013). *The origins and rise of dissident irish republicanism.* Bloomsbury Publishing USA.
94. Neumann, P.R. (2013). The trouble with radicalization. *International Affairs (Royal Institute of International Affairs 1944-),* [online] 89(4), pp.873–893. Available at: https://www.jstor.org/stable/23479398.
95. Nucifora, J., Kashima, Y., & Gallois, C. (2015). Influences on condom use among undergraduates: testing the theories of reasoned action and planned behaviour. pp.73–90. https://doi.org/10.4324/9780203769621-9.
96. Ogilvy, D. (1995). *Ogilvy on advertising.* Prion.
97. Pape, R. (2005). *Dying to Win.* Random House.
98. Parsons, T. (2013). Social system. *Routledge.* https://doi.org/10.4324/9780203992951
99. Petley, J., & Richardson, R. (Eds.). (2013). *Pointing the finger: Islam and Muslims in the British Media.* Oneworld Publications.
100. Picciolini, C. (2017). *White American Youth.* Hatchette Books.
101. Picciolini, C. (2020). *Breaking Hate.* Hachette Books.
102. Picciolini, C., & Jett, J. (2015). *Romantic violence: Memoirs of an American skinhead.* Goldmill Group.
103. Precht, T. (2007). *Home grown terrorism and Islamist radi- calisation in Europe.* [online] Available at: https://www.justitsministeriet.dk/sites/default/files/media/Arbejdsomraader/Forskning/Forskningspuljen/2011/2007/Home_grown_terrorism_and_Islamist_radicalisation_in_Europe_-_an_assessment_of_influencing_factors__2_.pdf.
104. Rambo, L. (1996). *Understanding religious conversion.* Yale University Press.
105. Reich, W. (1998). *Origins of terrorism: Psychologies, ideologies, theologies, states of mind.* Woodrow Wilson Center Press.
106. Sageman, M. (2008). *Leaderless Jihad Terror Networks in the Twenty-First Century.* University Of Pennsylvania Press.
107. Said, E. (1978). *Orientalism.* Vintage.
108. Schick, N. (2020). *Deep fakes and the infocalypse: What you urgently need to know.* Monoray.
109. Sedgwick, M. (2010). The concept of radicalization as a source of confusion. *Terrorism and Political Violence, 22*(4), 479–494. https://doi.org/10.1080/09546553.2010.491009
110. Shannon, C.E., & Weaver, W. (1949). The mathematical theory of communication. *The Mathematical Gazette,* [online] 34(310), p.312. https://doi.org/10.2307/3611062.
111. Silber, M.D., & Bhatt, A. (2007). *Radicalization in the west: The Homegrown Threat.* [online] Available at: http://www.nypdshield.org/public/SiteFiles/documents/nypd_report-radicalization_in_the_west.pdf.
112. Silke, A. (2004). *Research on terrorism: trends, achievements & failures.* Frank Cass.
113. Silke, A. (2010). *The psychology of counter-terrorism.* Routledge.
114. Storr, W. (2017). *Selfie.* Pan Macmillan.
115. Thomas, P. (2010). Failed and friendless: The UK's 'preventing violent extremism' programme. *The British Journal of Politics and International Relations, 12*(3), 442–458.
116. Torney-Purta, J. (2004). Adolescents' political socialization in changing contexts: An international study in the spirit of Nevitt Sanford. *Political Psychology,* [online] 25(3), pp.465–478. Available at: https://www.jstor.org/stable/3792552 [Accessed 24 Nov. 2022].
117. Ullmann, S., & Tomalin, M., eds (2023). *Counterspeech.* Taylor & Francis.
118. van Dyk, A. C. (2001). Traditional African1 beliefs and customs: Implications for AIDS education and prevention in Africa. *South African Journal of Psychology, 31*(2), 60–66. https://doi.org/10.1177/008124630103100208
119. Timothy Coombs, W., Falkheimer, J., Heide, M., & Young, P. (2017). *Strategic communication, social media and democracy: The challenge of the digital naturals.* Routledge.
120. Hallaq, W. (2018). *Restating orientalism a critique of modern knowledge.* Columbia University Press.

121. Weinstein, N. D. (1993). Testing four competing theories of health-protective behavior. *Health Psychology, 12*(4), 324–333. https://doi.org/10.1037/02786133124324
122. Zaidi, M. (2011). *The process of radicalization: contextualising the case of Pakistan.* [online] Pak Institute for Peace. Available at: https://www.pakpips.com/web/wp-content/uploads/2017/11/130.pdf [Accessed 25 May 2021].
123. Zuboff, S. (2018). *The age of surveillance capitalism: The fight for the future at the new fontier*. Basic Books.

Chapter 3
The Socialisation and Reasoned Action Strategic Communications' Model

Abstract This chapter introduces the "Socialisation and Reasoned Action Strategic Communications" (SoRaSCo) model as a conceptual framework to understand better and improve P/CVE (Preventing/Countering Violent Extremism) online communication efforts. Building upon the theoretical foundations presented in the preceding chapters, it integrates socialisation theory, strategic communications approaches, and behaviour change models—specifically the Reasoned Action Model (RAM)—to offer a more cohesive lens to examine radicalisation and counter-narratives impact. The chapter argues that traditional radicalisation paradigms, often rooted in linear and psychology-centric thinking, fail to fully capture extremist pathways' relational, social, and communicative dimensions. By contrast, SoRaSCo regards radicalisation as a dynamic social process shaped by the interplay of social agents, norms, and identity negotiations, all mediated by influential communicative platforms and narratives. The SoRaSCo model is then developed as a blueprint for more context-sensitive, audience-centred, and iterative P/CVE interventions. The approach highlights how strategic communications can counter extremist narratives more effectively through careful audience segmentation, ethically-informed data gathering, and continuous monitoring and evaluation. Ultimately, this chapter argues that integrating insights from socialisation, communication theory, and behaviour change frameworks is essential for designing and deploying counternarratives that resonate authentically and measurably reduce the appeal of extremist messages.

Keywords CVE—countering violent extremism · PVE—preventing violent extremism · Counter extremism · Communication · Strategic communication

Introduction

This chapter uses socialisation theory, behaviour change, and communications discourses to analyse P/CVE efforts and impacts. The three-pronged approach provides a distinct and powerful perspective that can be utilised to analyse radicalisation and extremist narratives. This chapter outlines the theoretical framework being

M. Zamir and C. Kaunert, *Strategic Communications' Role in Counter-Terrorism*,
Advanced Sciences and Technologies for Security Applications,
https://doi.org/10.1007/978-3-031-89110-6_3

deployed to investigate the impact of P/CVE work. Within this, globalisation trends are reviewed as critical influences on socialisation leading to radicalisation processes and subsequent approaches to deradicalisation. When considering the importance of communication in this particular context, it is crucial to comprehend the conceptual influence and rise of globalisation as a subset of socialisation theory. Globalisation is the social mechanism through which modern communication has had its most significant effect. Arguably, globalisation has had a significant socialising impact on how individuals adopt identities and communicate within those normative structures through the idea of human agency. Socialisation is a crucial learning process often enhanced by engaging in diverse social interactions. It enables individuals to acquire the appropriate attitudes, beliefs, skills, and information required for active involvement in a specific community or social group. In summary, integrating insights from socialisation and 'strategic communicative behaviour change' studies can develop a richer, more nuanced understanding of radicalisation. This could take into account its social and communicative dimensions. Such an approach could inform the development of more effective deradicalisation and CN strategies that target social ties, communicative techniques, and narratives that sustain radicalisation.

Consequently, studying radicalisation from this perspective necessitates understanding how this learning process functions in extremist environments. It enables us to investigate how socialising agents, such as radical clerics or online extremist communities, transmit extremist ideologies and how individuals are attracted, integrated, and retained in these groups. In addition, communication studies supplement this viewpoint by analysing the techniques, narratives, and symbols utilised to promote and justify extremist ideologies. Radical groups increasingly employ traditional and digital media platforms to disseminate their narratives. We can gain insight into the persuasive power of extremist narratives, their emotional and cognitive appeals, and how they create in-group solidarity while delegitimising and dehumanising out-groups by analysing the communicative strategies these groups employ. Efforts to undermine this process via P/CVE CNs can equally be analysed using the same conceptual approaches, which is the purpose of this book.

Traditional radicalisation theories, such as the push–pull and pathway models, emphasise macro-level socio-political factors, personal grievances, and psychological vulnerabilities as the primary drivers of radicalisation. While these approaches provide insightful information, they may overlook radicalisation's interpersonal, social, and communicative aspects. They also tend to portray radicalisation as a linear, deterministic process, a viewpoint that empirical studies are increasingly challenging. Alternatively, socialisation and communications models recognise radicalisation as a complex, multifaceted process entrenched within a social and communicative context. Individuals are no longer seen as passive recipients of extremist ideologies but as participants in a social learning process. In addition, examining the power of narratives and symbols using narrative and discourse analysis illuminates how extremist ideologies make sense to those who adopt them, their psychological appeal, and how they provide a sense of belonging, identity, and purpose. It also helps disclose the role of online platforms in facilitating radicalisation by offering spaces for the dissemination, reinforcement, and debate of extremist narratives.

As stated, the frameworks utilised in analysing P/CVE efforts incorporate socialisation theory to comprehend better the influences that contribute to radicalisation. Additionally, communications theory is employed to examine the impact of narratives on socialisation and the adoption of beliefs. Lastly, behavioural change theories enable us to understand how narratives can shape individuals' actions following their socialised ideologies. Below, this chapter outlines the key benefits of the chosen approach, followed by critical limitations, and then looks at developmentally applying the model to the research aims.

Benefits of Adopting a Socialisation-Based Communications Theory Framework

Socialisation theory offers dynamic (interactional), structural and cultural understandings of how people develop and learn values to participate in society. In utilising socialisation theory as a foundational element for this book, the overall benefits of the approach can be distilled into different but interlinked categories. These categories are pivotal in addressing the conceptual and practical processes required to understand the extremist pathway and individual nuances. Below is an explanation of the perceived benefits of socialisation-based theory over the currently favoured radicalisation discourse-based approaches within P/CVE.

Comprehensive Understanding: Socialisation and communication models enhance our understanding of radicalisation by addressing the what, the why, and the how. They focus on the learning and communication processes involved in individuals adopting extremist ideologies.

Understanding of Social Dynamics: This method permits the examination of the social dynamics of radical groups and online communities. It reveals how these organisations operate, how members interact with one another, and how social relationships within these environments facilitate radicalisation.

Focus on Narratives and Symbolism: The emphasis on narratives and symbolism offers a nuanced comprehension of the persuasive power of extremist ideologies. It illuminates the emotive and cognitive appeals of extremist narratives, their role in group identity formation, and how they delegitimise and dehumanise outgroups.

These perceived benefits are also supported and expanded through the links between socialisation theory and communication-based behaviour change via considering.

Role of Communication Platforms: This approach emphasises the importance of communication platforms, particularly digital ones, in propagating and reinforcing extremist ideologies. It focuses on how these platforms facilitate the dissemination, social reinforcement, and discussion of extremist narratives.

Targeted Interventions: The insights obtained from this methodology can be applied to developing more efficient interventions. By understanding the social and communicative aspects of radicalisation, interventions can be better tailored to disrupt these processes, challenge extremist narratives, and offer alternatives that meet the social and psychological needs exploited by extremist groups.

Individual Agency: Contrary to some now 'traditional' models of radicalisation, the socialisation and communication approach acknowledges individuals' active role in their radicalisation process. This perspective can aid in developing more respectful and efficient engagement strategies for deradicalisation initiatives. However, whilst socialisation theory does appear to provide both nuance, depth, and greater emphasis on the role of social processes, the approach is not without its limitations, the most relevant of which are outlined below.

Criticisms of Socialisation-Based Approaches

While this model offers a great deal of adaptative potential and uses interlinked disciplines and concepts, it has some relevant limitations. These are summarised below and will be used to build an improved overarching model for this book and, hopefully, future research.

Over-emphasis on Social Aspects: Critics suggest that this model may over-emphasize the role of social learning and communication, thereby downplaying other significant factors such as personal grievances, psychological factors, socio-economic conditions, and geopolitical issues.

Inability to Isolate Variables: given radicalisation's complexity and multifaceted character, isolating socialisation and communication as distinct components of the process can be challenging. This difficulty can impede the development of targeted interventions and confound the interpretation of findings. Contrary to other radicalisation models, the socialisation and communication approach places less emphasis on anticipating who will become radicalised. Its primary focus on comprehending radicalisation processes may limit its usefulness for forecasting or preventative efforts.

Self-Reporting: Most of the evidence for social learning and communication processes comes from interviews or self-reporting by radicalised individuals. However, due to the subjectivity of individual perceptions or the possibility of manipulation, this reliance on self-reporting may result in bias or inaccuracy.

Data Privacy and Ethical Concerns: effective research involving digital communication platforms frequently necessitates large quantities of data, which raises ethical concerns. Furthermore, engaging with extremist narratives for academic purposes can present ethical challenges. These limitations are also linked to broader aspects of social and political discourse, such as;

Overlooking the Role of State Actors: This approach frequently focuses on non-state actors. It may disregard the role of state actors or institutions in fostering extremism, especially in regions experiencing conflict or stringent state security measures.

Despite these criticisms, socialisation and communication studies models remain valuable approaches that complement other theories and models in the field by contributing to a broader, multifaceted understanding of radicalisation. Thus, these models require development work, which would significantly enhance both their place and usage within P/CVE research. In this regard, there are core areas of development that could be suggested to achieve this objective.

Developing and Improving the Socialisation and Communication Model Within P/CVE

P/CVE programmes seek to deter individuals from engaging in radical, politically, socially, culturally, or religiously motivated violence. Comprehending and creating effective CNs are essential components of these efforts. A key element within this sector is using behaviour change models to achieve measurable change. The socialisation and communication model could be enhanced with behaviour change needs in the context of P/CVE in the following ways:

The Integration of Disciplines: Combining the socialisation and communication model with insights from psychology, political science, and sociology could yield a more comprehensive understanding of radicalisation. This would facilitate the creation of CNs that address the social and communicative aspects of extremism and the personal, political, and socioeconomic factors at play.

Cultural and Contextual Awareness: Radicalisation processes and the efficacy of CNs can vary considerably depending on cultural and local contexts. Consequently, the model could be improved by incorporating more robust mechanisms for considering these factors, ensuring that CNs are pertinent, persuasive, and respectful of cultural differences.

User-Centred Design: CNs can be more effective if created with a thorough comprehension of the audience's motivations, requirements, and communication patterns. Consequently, user-centred research techniques from disciplines such as marketing and user experience design could be integrated into the model.

Evaluation and Iteration: It is crucial to develop mechanisms for rigorously testing and evaluating the efficacy of CNs. This includes quantitative measurements (such as shifts in attitudes or behaviours) and qualitative feedback. These evaluations ought to be utilised to iterate and enhance CNs perpetually.

Coordination with Technological Platforms: Given the role of online platforms in radicalisation, it may be advantageous to collaborate more closely with these platforms. This may involve sharing anonymised data for research, developing collaborative initiatives to combat extremism, or creating guidelines for dealing with extremist content.

Development of Ethical Guidelines: given the potential privacy and ethical concerns associated with studying extremist narratives and online communities, more straightforward ethical guidelines could be developed. This could involve input from many stakeholders, including academics, policymakers, tech corporations, and civil society representatives. By incorporating these enhancements, the socialisation and communication model could become an even more potent instrument for comprehending radicalisation and creating effective CNs in the context of P/CVE. In seeking to develop a more detailed understanding of this potential model, the following sections build on socialisation, communications and behaviour change approaches. Specifically, the acquisition of norms and values via socialisation is a crucial element of the model, with approaches to strategic communications and the RAM of behaviour change providing a segue from theory to operationalisation modelling for this research effort. As such, the latter two elements will be introduced as sub-elements of an overarching socialisation-based model for this book.

SoRaSCo—'Socialisation and Reasoned Action Strategic Communications' Model

Below is a revised socialisation and communications model that could be used to design strategic P/CVE CN communication. This model could be further researched and built upon following changes to tech and communications, such as Machine Learning Language models, changing geopolitical landscapes, and shifts in social dynamics, both online and offline. The key deployable elements of the model would be Research and Contextual Knowledge, CN Development and Testing, Platform Approach, Cooperation and Advocacy, Monitoring and Evaluation, and Iteration and Enhancement.

Research and Contextual Knowledge: The initial stage entails extensive research and comprehension of the radicalisation process, the extremist narratives, and the intended audience. This includes the study of the social dynamics within extremist communities, the symbolic and emotive appeal of their narratives, their communication strategies, and the local, cultural, and personal factors that may influence an individual's susceptibility to these narratives. Utilise interdisciplinary perspectives to conduct a thorough investigation. This is precisely where socialisation theory and (strategic) communications approaches can provide the element of nuance and depth of field often missing from rushed and 'quick-win-based communications efforts. Audience segmentation can be crucial for effective CN strategies, just as in marketing campaigns. Segment the audience based on age, radicalisation level, socio-economic status, geographical location, psychological profile, online behaviour, and specific grievances or motivations.

CN Development and Testing: Develop and test CNs that directly respond to the appeals of extremist narratives. These should provide alternatives to the grievances or needs exploited by extremist groups, promote tolerant and inclusive values, and debunk misinformation or propaganda. Using language, symbols, and practical media formats for each segment should be designed to resonate with the specific target segments.

Platform Approach: Determine the most effective platforms for disseminating CNs, considering where the target audience segments are most active and where extremist narratives are spreading. This may involve social media platforms, online forums, messaging applications, or offline channels.

Cooperation and Advocacy: Collaborate with tech platforms, local communities, influencers, and other stakeholders who can help amplify CNs and advocate for their promoted values.

Monitoring and Evaluation: Implement mechanisms for monitoring the dissemination of CNs and assessing their influence. This should include quantitative measurements (such as reach, participation, and behaviour change) and qualitative feedback.

Iteration and Enhancement: Utilise the evaluation's findings to continuously enhance the CNs and adapt the strategy to changing conditions. Ensure precise adherence to ethical guidelines throughout all these steps to respect privacy, avoid harm, and preserve the credibility and integrity of the CN campaign.

It can be argued that this model differs from much of the current understanding within operational P/CVE contexts in the following ways:

The proposed framework of holistic integration combines two fundamental viewpoints: socialisation and communication studies. These often need to be comprehensively integrated into existing approaches to P/CVE. By integrating these viewpoints, a comprehensive understanding of radicalisation is achieved, encompassing the intricate relationship between social processes, communication dynamics, interpretive social agency, and social action in developing extremist ideologies. These elements could take shape and form in the design and development context.

The communicative socialisation paradigm clearly emphasises the importance of learning processes, social interactions, and communication channels within extremist environments. This distinguishes it from other current tactics in P/CVE that also recognise the impact of social factors. This book investigates how individuals acquire and internalise extreme views through social contact.

The user-centred design framework suggests using audience segmentation and user-centred research to create CNs customised for distinct target groups. This approach surpasses a uniform strategy and acknowledges the significance of effectively appealing to varied groups in the fight against extremist messages.

The SoRaSCo model acknowledges the significance of online platforms in facilitating the dissemination of extremist beliefs within digital radicalisation. The approach argues for collaborating with tech platforms and implementing a platform strategy that carefully evaluates the most efficient methods for distributing CNs.

The framework facilitates a cyclical and iterative process, enabling the ongoing enhancement of CN initiatives through systematically monitoring and evaluating their effectiveness. This highlights the importance of adjusting to evolving circumstances, new patterns, and input from target demographics.

Ethical considerations have long been integral to P/CVE initiatives. However, the SoRaSCo framework emphasises ethical adherence at every stage of the process. This includes ensuring data privacy, preventing harm, and upholding the credibility of the CN campaign.

The framework seeks to overcome the existing constraints in P/CVE by integrating these components. It endeavours to provide a more sophisticated, flexible, and responsive strategy for countering radicalisation and fostering positive narratives among various target groups. The need to address both the social and communicative aspects of extremist narratives is underscored, leading to the development of more impactful and contextually appropriate tactics for P/CVE. Behavioural change elements in this guise take on more meaningful features. This is achieved by adopting and utilising user-based engagement in both design and delivery, thus enhancing the nuance and suitability of specific messaging themes to better match audiences' needs.

At the heart of both the conceptual and applied assumptions underlying efforts to dissuade, discourage, and eventually 'positively' influence 'susceptible' or 'radicalised' persons away from the possibility of violence is the frequently overlooked idea of socialisation. In terms of comprehending (1) how to approach an in-depth study of CNs within P/CVE conceptually and (2) how to posit evaluative and developmental ideas for future efforts, the requirement to source and access an adequate theoretical framework results in socialisation and its development within psychology, political science, and sociology appearing as both an epistemological conduit and an ontological outcome of various approaches in the field [6, 7, 9]. The reason for this dual-purpose application of the idea is straightforward to explain but very hard to superimpose on the immense subtlety and frequently opaque discourses of radicalisation and deradicalisation [2]. The SoRaSCo model is intended to be an iterative, cyclical process instead of a linear one. Radicalisation is dynamic and complex, so effective CN strategies must also be dynamic and adaptable. For this research book, it is hoped the approach adopted and its benefits/limitations have been adequately explained along with a detailed look at the respective models in their context. The following section builds on this exploration by identifying suitable methodological means by which this book can understand, gather, and determine the processes behind strategic communications-based P/CVE efforts.

Significance and Originality of the Research

There have been several attempts to explain and analyse CN efforts at the macro level of P/CVE measurement and evaluation processes. These include government, private donor-funded think tanks, and EU-wide practitioner-led networks. Such efforts have sought to shed light on the process, position, and potency of CN efforts [12]. In general, and by way of surmising these, they have often taken the form of 'toolkits' and 'good-practice' guides that offer top-line descriptions of the 'what', 'how', and 'to whom' elements of strategic communications and practice from a general viewpoint. To explain this point, 'what' tends to include the name and type of CN (website, leaflet, video, testimony, etc.), 'how' tends to explain the medium or dissemination choices of the effort and looks at the technical style or narrative component, i.e. CN or alternative narrative, counter-speech or alike. The 'to whom' offers a basic rendering of a 'target audience' and often uses terms like 'at-risk' or 'vulnerable' [15].

The trouble is that while the framework for doing this appears useful, the application needs to address the same issue as mentioned earlier, in that minimal mention of any theoretical assumptions, conceptual design considerations, narrative choice discussions and outcome effects is found. In short, this suggests that the 'strategy' for this endeavour, somewhat ironically, lacks a robust strategy. As a result, we find ourselves reading brochures about 'resounding success', 'audience segmentation' and 'impactful reach' without any critical analysis of why and how these claims can be made or substantiated. This research book seeks to build and embellish robust conceptual frames for studying P/CVE CNs and their strategic intent.

There is little or no context-specific foundational knowledge on which to base these claims. This is why the field of CN study posits knowledge in the 'post-effect' context and not in the 'pre-effect' context (theory and design stage). Using the SoRaSCo model in this study, the approach and insights presented offer a unique and original look at the P/CVE sector's communications efforts. Through detailed narrative analysis using case studies (Chap. 4 and 5) and an overarching analysis of the role of strategic communication within P/CVE in Chap. 6, a clear gap in current understanding will hopefully be filled. Without seeking to be too offensive to these efforts, it does appear akin to the notion of one party seeking an answer from another regarding how their headache 'was controlled', and the answer comes back as 'the taking of a pain killer'. The manifestation of the pain (headache) and supposed relief (pain killer) are not examined from a diagnostic, formative, and summative perspective regarding what caused the build-up of pain and how the remedial action affected the pain and may have also created unintended side effects. Too often, there is confusion between correlation, meaning causality, from an evaluation and measurement viewpoint [19]. The 'art' of the CN has been studied through a dubious scientific lens of enquiry in various ways but without any detailed consideration of the environmental and contextual contexts in which these activities existed. In other words, the analysis is often too far removed from the environment of the source material to be helpful beyond a summative/descriptive account of 'what we think happened and why'.

This raises questions about how epistemological and ontological considerations are managed within this work. It is acknowledged, however, that this effort has pushed the body of knowledge in this field towards the next stage of conceptual and evaluative endeavour, where the need is for a multifaceted and detailed study into the underlying assumptions of the terminology, theoretical assumptions, cultural context, ideological context, and effects/impact measurements used in CN efforts that are based on as holistic an assessment as is possible within the constraints of an 'information shy' P/CVE sector. This current research effort seeks to do this and, in doing so, to push the learning dial further towards a fuller picture regarding CNs and their perceived need, use, and effectiveness.

In terms of discourses on radicalisation, deradicalisation, and counter-radicalisation efforts through CNs, only Kurt Braddock's concept of 'weaponised words' has attempted to synthesise theoretical and applied considerations in the field using RAM and the concept of persuasion as the conduit between narrative and behaviour [2]. This book aims to build on previous research by delving deeper into the construct, creative design, strategy, communication methods, behaviour change assertions, audience targeting, dissemination, and impact measurements of CN efforts to better understand the field's barriers, moral choices, stakeholder contexts, and social impact. In doing so, the hope is to reveal the risk, management, and measurement factors that pervade this subject, improving the effectiveness of challenging extremist rhetoric in the future.

The previous chapter presented several anomalous patterns in contemporary understandings of radicalisation studies and associated CN efforts. The primary concerns stem from radicalisation as a vehicle for deradicalisation communication without rigorous intellectual and theoretical analysis of the concepts and their application. Although socialisation theory and political socialisation were widespread conceptual paradigms before 9/11, they have been pushed aside in favour of an idea (radicalisation) that lacks depth regarding nuance and flux in human behaviour and, somewhat frustratingly, also lacks empirical reliability when suggesting narrower and more refined understandings of its auspices.

The sheer volume of CN efforts that have permeated the online space over the last decade—from government agencies to creative agencies and those 'actors' in between—necessitates thoroughly examining their conceptual underpinnings and potential or measurable impact. From an interpretative, empirical, and illustrative standpoint, this is a critical and apparent gap in the field. While this research will not attempt to quantify the findings empirically, the research parameters allow for the examination and interpretation of existing, available, and empirical data using the theoretical framework outlined above. This will enable the findings to contribute to the development of empirical questions and gaps in the field's current approach through detailed analysis and critique of efforts perceived to be successful and those perceived to be weak.

By utilising the theoretical framework and applying it to all case studies, as well as conducting an in-depth discourse and gap analysis, the key findings are expected to present a rich picture of lessons learned, future needs, and development issues. At the same time, it appears imprudent to suggest that the majority of CN efforts to date have taken an uncoordinated approach (no particular organised or strategic plan regarding whom to target, with what message, and how) to strategy, design, and evaluation. This is, unfortunately, something that is also levied at CT policy broadly. Critics argue that too many attempts to deliver CT policy at the operational level have been based on the need for 'quick wins' and a lack of rigorous planning or evidential base. There is a need in this field to advance understanding of, and distinguish between, strategy and tactics. If good practice and sound conceptual ideas are prevalent in certain CN efforts, these must be synthesised to enable governments, the tech sector, and civil society to work more in tandem based on common understandings of the problem sets and solutions offered by strategic communications.

CN and CT efforts more broadly carry with them the distinct impediment of stringent information management and sharing control for security-centric reasons. In the case of CT, this context is often explained from a national security perspective, citing intelligence-oriented considerations as to why raw data is unsuitable for sharing in the public domain [21]. CT operates in various guises, from active law enforcement and intelligence-led operations to critical infrastructure and crisis planning [17]. When the matter concerns stopping potential acts of terrorism or apprehending individuals suspected of initiating such acts, the flow of information to public bodies and mainstream media is often sparse and managed in compassionate and structurally process-driven ways [22].

The overarching picture is that this field is closely allied to government policy and operational responses to an 'existential level of threat', assessed using risk and probability indicators [14]. This continues to be a significant drawback for research in this area. One of the possible avenues to alleviate this strain on information flow has been the development of embedded academia and policy think tanks closely allied to state funding, which actively research terrorism-related matters alongside the state as a 'principal stakeholder' [16]. Sometimes, data and information are processed, and grant research partners have access only to the processed versions. In other instances, initiatives and campaigns that are financially supported by the government are subject to measurement and evaluation at the hands of centralised departments, which delegate data collection down to the municipal level [4]. CN efforts are somewhat convoluted in this regard and are riddled with an element of mystery regarding sponsorship, funding, in-kind support, and attribution [10]. From a specifically preventative context, this ambiguous relationship warrants closer attention and explanation if the question of the effectiveness of such efforts is to be an indicator of any groundswell of positive change in the battle of narratives, ideologies, and social systems [8].

This book seeks to understand the world of P/CVE CNs at a very granular level. We know that CNs exist, and they permeate the public realm of the internet and social media. In other words, they reach audiences in some way, shape, or form. The aim here is to unravel the role of strategic communications in such efforts and ultimately identify some framework for establishing how counternarratives can be seen as effective or ineffective and which elements of current pedagogy can be classified as intuitive or counter-intuitive. The immediate problem a researcher faces in this context is relatively obvious and has already been alluded to: access to information. Moreover, any existing information is a secondary source at best and heavily redacted to protect various stakeholders' and partners' positions [2]. Proponents of CN efforts are not queuing up to reveal their methods, identities, partnerships, and ideological positions. This means that any effort to measure and build a body of evidence at the nexus between the practitioner and academic enquiry of CN P/CVE communications, beyond the current practice of essential overview-level reviews of such efforts by non-governmental organisations (herein NGOs), think tanks, and other external entities, is likely to be the first of its kind in the research environment. In choosing the title and scope of this research book, in-depth considerations have been given to the state of the field, the theoretical assumptions behind this work, and the absence of wholly innovative cross-discipline analysis and discussion on the ultimate purpose of such efforts. The field of CNs is arguably, therefore, a unique facet of P/CVE because not only does it exist at the point of 'theory into practice' or vice versa, but it also contains the most diverse range of theoretical, policy, social, cultural, political, and communicative assumptions regarding the underlying efficacy of the work [10, 20].

Research Design and Method

Initial scoping, review, and fieldwork have revealed a central theme of a (previously mentioned) somewhat 'scatter-gun' approach to P/CVE CNs regarding their conception and need (purpose), alongside some essential development of an overall framework within which these efforts are loosely categorised. This framework helps to contextualise the P/CVE online CN space to some extent. Still, it has not been critically examined in detail regarding its conceptual suitability for strategic communications-based P/CVE efforts. The framework in question has two distinct roots concerning the knowledge base used to apply this to P/CVE efforts. One element arises from but appears also to have been adapted from harm-reduction and safeguarding models used predominantly in gang prevention and related anti-social behaviour risk mitigation efforts [11]. It utilises a three-tier pyramid that places universal vulnerabilities at the lowest level, specific at the middle level and highly acute vulnerabilities at the top [15]. The vulnerabilities referenced in this context are vulnerability to radicalisation, radicalised, and propensity to violence through radicalisation. A second element is based on behaviour change models, such as the RAM, which amalgamates the TRA and TPB into one overarching framework. The intention is to analyse these models in light of the literature review findings and theory framework presented to assess possible gaps in understanding and what assumptions may have created these when applied to strategic communications efforts in P/CVE. By taking socialisation theory as the base for this book, and adding strategic communications and the behaviour change theory of the RAM as support layers, this book seeks to:

1. Study and analyse the strategic design, content/discourse, and delivery tools to understand the effectiveness, limitations, and learning from counter-radicalisation/CN campaigns/interventions and inherent behaviour change assumptions contained within and
2. Assess the scope and opportunities for more effective P/CVE online efforts based on emerging tech tools and more innovative approaches to understanding problem sets.

The book will examine the P/CVE CN story from its nascent emergence approximately a decade ago to more recent iterations and efforts to comprehensively study this field. The lack of public access to date to such efforts means that the choice of example campaigns can be construed as limited. Still, equally, the chosen campaigns for this book represent the most trail-blazing and pioneering efforts to combat radicalisation and extremism in the online space. The choice had to be made based on suitability to assess the pyramid model and its use and available information on the campaigns. The three campaigns to be studied are:

1. Abdullah-X (YouTube/Google/EU/RAN/CSO campaign)
2. One2One/Counter Conversations—phase one and two (Facebook/Meta), ISD)
3. Search Redirect (Google)

Abdullah-X (AX) is among the most innovative and impactful P/CVE online efforts [2]. It is seen as a key example of a then-nascent CN domain fuelled by a fear of thousands of Western Muslim youths seemingly determined to join Daesh and an urgent need to counter this fear with some form of strategic communications effort. The Counter Conversations campaign (initially called One2One but referred to as Counter Conversations or CC for simplicity throughout this book) is the best example of the role of technology and tech tools, as well as how policy change can impact P/CVE efforts, including the issues of risk and harm for more deceptive actors who may use the same tools for nefarious means. Search Redirect (herein SR) serves as an example of online efforts based on user behaviour and tools to detect and redirect users to alternative content. These examples provide sufficient depth of field to assess the P/CVE CN domain in various guises and contexts, aiming to examine impact, gaps, and learning concerning the effectiveness of behaviour change-based approaches.

The research method chosen utilises case studies alongside narrative and discourse analysis. This allows for a greater focus on addressing these campaigns' conceptual, strategic, and impact issues, as well as related policy issues and VEO trends in messaging and recruitment. As such, the main component of the research will be qualitative and interpretive in focus. This method will assist in aligning with the book's broad hypothetical assumptions about the nature of radicalisation being intrinsically linked to the process of socialisation and the idea that the current P/CVE understanding of CNs is counter-intuitive to addressing deradicalisation needs in their own right substantively. The COVID-19 pandemic disrupted many original research goals at the time of the book's initiation. Part of the original methodological approach would have included a combination of quantitative analysis to build a comprehensive picture of the online CN field and qualitative tools such as narrative analysis and case studies for exploratory, descriptive, and explanatory purposes. For many online campaigns to be included in this fashion, the quantitative element requires approved access to campaign data, stakeholders, creators, and, ideally, end-users. Attempts to interact with a multitude of global campaigns, NGOs, CSOs, and government officials, as well as their commercial partners, became an uphill battle,

Upon reflection, such a notion was outside the realm of this context of investigation and the resources available. To combat this limitation, an additional layer of interpretive analysis will be utilised to harness learnings and processes associated with offline P/CVE interventions. This will enable a detailed comparative study of the frameworks mentioned in earlier chapters by focusing on prevention/harm reduction approaches of offline interventions and offsetting these with the more overt version of behaviour change-based intervention used in online CNs.

Of course, it is fair to point out that, at their best, P/CVE CNs would include elements of both frameworks or approaches considered 'effective' to a greater degree. For this book, however, and based on limitations already outlined, the prevention/

harm-reduction approach and offline interventions offer a foundational comparator for the newer and more cerebral online P/CVE CN field, which relies heavily on persuasive communication and behaviour change-based design elements.

The key objectives of this research book are to develop a critical discourse analysis of both the notion and actuality of the 'CN campaign' approach through case study analysis, examining the state of the craft in CN communication from a multi-layered perspective (behaviour change, prevention, and persuasion). The overall goal is to provide a comprehensive examination of the tools, techniques, and conceptual underpinnings of CN efforts to date, including the AX campaign, which was named 'global best practice' at the 2015 Global P/CVE summit in Washington, DC, and has since spawned several other campaigns seeking to capitalise on the 'lessons learned' from this pioneering work [15]. The CC and SR campaigns are both innovative and impactful examples of the role of technology and public–private partnerships as means to counter radicalisation. All three campaigns need to be studied for the assumptions under which they were formulated and disseminated. As mentioned earlier, these examples offer a chance to assess conceptual ideas around broad-brush or mass-audience versions of CNs and more specific targeted iterations, which fit nicely into studying the tiered pyramid model of interventions in the P/CVE space.

The AX campaign, in particular, was unique in design and execution, attracting the tech sector, non-governmental organisations, government, media, and the global practitioner community for 'lessons learned' and critical insights that could benefit future planning and design for such endeavours. The researcher's fieldwork and access to the inner workings of this campaign offer a chance to detail, in as much depth as possible, how this campaign paved the way for the learning and development of most subsequent P/CVE online efforts, including the knowledge base that went to CC and SR, respectively. Given that the campaign and project formally concluded in late 2017 in terms of new content transmission, and given the passage of time and the build-out of the field since then, the opportunity to unearth the processes behind such work is now achievable.

The rationale for this assertion is straightforward: for years, the AX campaign and project have been employed in practitioner-academic contexts without holistic appraisal of the methodology, procedures, and conceptual ideas underlying the work. In other words, up to this point, only publicly available information has been used to determine the project's objectives, parameters, and mode of operation. The AX project's legacy will be addressed in this book by applying the P/CVE field's assumptions at the time, current thinking in the field, and then laying the groundwork for the P/CVE CN paradigm to improve its comprehension moving forward. The goal is to examine how the CN sector operates and its linkages to broader CT issues. This should help to advance the discourse of P/CVE toward greater robustness and relevance in a shifting global context for violent extremism and the narratives that support it.

Considering the abovementioned restrictions, the suggested study model would employ a mixed-methods approach within the qualitative realm of social science-based research. This technique is provocative because it strikes at the heart of current arguments regarding the state and nature of social scientific research [5]. This is connected to the academic and philosophical challenge posed to positivist social science foundations by a prevalent and evolving interpretivist approach, exemplified perhaps most dramatically by the emergence of the 'critical studies' field of research, post-modernist theory, and the post-truth narrative, which are currently challenging every assumption that social sciences have made since the Vienna Circle meetings of the 1920s. This book illustrates this ongoing argument and supposition through the approach to research design and methodological considerations in utilising case studies. Traditionally, the application and design of research methodologies have been viewed as hierarchical processes that involve a system of ordering and organisation. Indeed, one may argue that many social scientists regard case studies as suitable only during the exploratory phase of a research enquiry. However, one could contend that employing such methodologies has always been an integral component of the social scientific approach, with any experiment being fundamentally exploratory [18]. This is further supported by evidence from social research of case studies that move beyond the exploratory to explanatory domains and employ descriptive/interpretive tools [1]. As Lin points out, a more pluralistic view is that "any research method can be employed for all three purposes—exploratory, descriptive, and explanatory" [18].

It is worth noting here that specific parameters must be considered when terms like 'what', 'who', and 'why' are employed in a research question. Two alternatives exist when the term 'what' is used. Suppose the topic is 'What lessons may be learned from P/CVE CN efforts in the internet space?'. In that case, it may allude to some exploratory inquiry, as the question contains adequate rationale for its necessity. The primary reason for this position is that the subject might be addressed using various research approaches. These may include surveys, comparative benefits of one model over another, or case studies examining various contextual elements in a particular field [13]. For the second type of 'what' question, connected to a series of subquestions such as 'to what extent', archival approaches may be the most effective because they involve contextual and historical source analysis and inquiry.

In comparison, if the research topic involves the words 'how' or 'why', these are typically the areas of enquiry that benefit from utilising case studies, historical research, and even experiments. Lin believes this is because these queries examine operational processes over fixed or open time periods and are not explicitly concerned with incidence or frequency-based anomalies [18]. In many instances, it can also be more of a process of deciphering what the title of the research question leans itself towards in terms of a possible research method. From a substantive perspective, methods can overlap, and this is perhaps the key to understanding both the design element and the actual study process [3]. In the case of the study being conducted, the central question is built around the role of strategic communications with P/CVE CNs and what lessons can be learnt from such efforts.

Ragin and Becker argue that despite its extensive use and relevance in social science discourse, social research does not fully define the term 'case' and the many terminologies associated with case analysis. Implicit in most social science concepts of case analysis is the notion that the objects of enquiry are sufficiently similar and distinct to be treated as comparable instances of the same general phenomenon. At the very least, most social sciences feel that their methods are sufficiently robust to overcome the inherent uniqueness of objects and occurrences in the social world. Case study social scientists argue that their cases are representative, exemplary, extreme, or conceptually decisive in some way. Thus, even in case study research, assertions about the relationship between the chosen instance and other cases frequently involve repetition (Ragin and [19].

In some ways, this current research seeks to unearth elements of conceptual uniqueness and identify the parameters and factors that suggest one type of CN effort may be more intuitive than another under certain conditions. Again, however, there is an argument at the macro level that while it is easy to see case studies as a subset of qualitative analysis and even to equate the two, practically every social science study is or may be conceptualised as a case study, frequently from a range of perspectives. At the absolute least, each research study is a 'case study' because it examines social phenomena unique to a particular period and place [19].

Analysis Criteria—RAM, Socialisation Theory and Communications Approaches/Contexts

Using the RAM as a framework for analysing behaviour change in case studies related to P/CVE strategic communications entails a comprehensive evaluation of distinct operational components while integrating pertinent theories from socialisation and communication studies as the 'communicative socialisation model' outlined in Chap. 2. This model is applied using a qualitative methodological approach outlined in this chapter, leaning on discourse analysis to provide a detailed account of the project's history. To enable the chosen model to be applied effectively, the case study begins with a detailed discourse analysis, setting the scene for the application of the model to follow.

This approach aims to comprehensively overview the fundamental components and their interconnections with RAM, socialisation theory, and strategic communications studies. The criteria for analysis of the case studies use the RAM approach's central features, of which there are five in total (listed below) and underpins them with a discussion of the socialisation-based considerations and communications context where relevant. These are then located alongside the theory framework model of SoRaSCo outlined in Chap. 2, which also emphasises the need for effective feedback and iterative approaches to P/CVE CN communication design. Beyond the discourse analysis, this offers six additional features and criteria to analyse where relevant, listed below. Following this list, the criteria are detailed, including socialisation and communications considerations, which have been examined within the case studies where applicable.

1. Behavioural Intention
2. Attitude Towards Behaviour
3. Subjective Norms

4. Perceived Behavioural Control
5. Environmental/Structural/Background Contextual Factors
6. Feedback and Iterative Approaches (additional element for a RAM+ approach)

RAM and Behavioural Intention

Behavioural intention is a fundamental focus of the RAM framework, as it is considered a primary factor in determining behaviour. Within the realm of P/CVE strategic communications, the process includes examining and assessing the target audience's inclinations towards engaging in violent extremism, as well as identifying various elements that influence these inclinations.

Socialisation Context

It investigates how various socialisation agents, including family, peers, and media, influence individuals' ideas, attitudes, and intentions. This study examines the potential effects of exposure to extremist narratives or CNs inside various socialisation agents on individuals' intents.

Communications Context

It aims to analyse the efficacy of various communication methods in influencing behavioural intentions. Specifically, this analysis will focus on the impact of persuasive messages, narratives, and communication channels (such as social media and community participation) on moulding individuals' behavioural intentions.

RAM and Attitudes Towards Behaviour

RAM values attitudes as a crucial determinant of behavioural intention. Examining views about violent extremism, peace, tolerance, and inclusivity is essential within the framework of P/CVE.

Socialisation Context

Specifically, it focuses on the processes by which attitudes towards violence and extremism are shaped, strengthened, or contested as a result of encounters with various socialisation agents. This is designed to comprehend better the significance of cognitive and affective mechanisms in developing attitudes.

Communications Context

It aims to evaluate the influence of diverse messaging and narrative framing strategies on modifying attitudes towards violent extremism within communication studies. This assessment will consider several elements, including cognitive dissonance and emotional appeals, to understand the effects of these communication techniques comprehensively.

RAM and Subjective Norms

Subjective norms, as a component of the RAM framework, encompass an individual's sense of the societal pressures exerted upon them to partake in a specific behaviour. The field of P/CVE includes the examination of how social norms prevalent within a given society influence the degree of acceptance or rejection of violent extremist beliefs.

Socialisation Context

This criterion explores the impact of social networks on individuals and how prevailing norms and values impact them. Additionally, this analysis investigates how conformity or non-compliance to extremist ideologies is reinforced inside these social networks.

Communications Context

It aims to explore the influence and impact of influencers and opinion leaders on forming and disseminating subjective norms within the field of communication studies. This analysis seeks to assess the efficacy of community-based strategies and testimony former extremists provide in challenging established societal norms.

RAM and Perceived Behavioural Control

Perceived behavioural control (PBC) is a construct considered in the RAM. PBC pertains to an individual's subjective view of their capability to engage in a specific behaviour. Analysing perceived barriers and facilitators to disengagement from violent extremism is a crucial aspect of P/CVE.

Socialisation Context

Investigate the impact of socialisation events on individuals' self-efficacy and perceived control over their beliefs and actions, as proposed by socialisation theory.

Communication Context

This study seeks to assess the effects of communication methods that prioritise empowerment. These tactics are designed to bolster individuals' PBC by offering them tools and support to disengage from certain behaviours or situations.

RAM and Environmental/Structural/Background Contextual Factors

RAM sees environmental and structural elements as significant influences on behaviour. Within the field of P/CVE, an essential aspect is the examination and assessment of contextual elements, such as political, economic, and social circumstances, which can contribute to the attractiveness and allure of violent extremism.

Socialisation Context

It aims to examine the impact of macro-level factors on forming individual beliefs and attitudes and to explore how these factors interact with micro-level influences.

Communications Context

It aims to evaluate the efficacy of strategic communications initiatives in tackling the underlying causes and structural factors contributing to violent extremism.

RAM, Feedback and Iterative Approaches

Additionally, a feedback and iterative approach will be employed to enhance the success of these efforts. RAM emphasises the importance of maintaining a continual feedback loop to improve interventions. In the realm of P/CVE strategic communications, it is imperative to consistently evaluate the efficacy of messaging and, after that, modify the approach following received input.

Socialisation Context

Gains insight into how individuals' perceptions and behavioural intentions are influenced by their encounters with counter-messaging and disengagement efforts.

Communications Context

Employs audience research and message testing to continuously enhance communication strategies and guarantee alignment with the intended target audience.

By incorporating the RAM approach with socialisation theory and communication studies, strategic communications efforts in P/CVE can thoroughly comprehend the factors that motivate violent extremism. Consequently, this integration can facilitate the development of more efficacious interventions to modify extremist behaviour. This methodology promotes a comprehensive examination that considers personal beliefs and attitudes, societal influences, methods of communication, and broader environmental elements that contribute to the modification of behaviour within the framework of combating violent extremism. This assertion will be applied and tested in the following case study discussions. It is essential to acknowledge that these criteria for analysis are, on the one hand, specific and tailored to the RAM approach, but on the other hand, broad in range and, therefore, may or may not apply equally to the case studies when analysed. This exercise will benefit the P/CVE field regardless, given that gap analysis and suitability needs are essential for constant improvement.

The case studies used in this research project have been selected based on the following set of features:

- Delivery of content/intervention in the online space
- Use of CN tactics
- Application of creative or innovative methods.

The first case study (AX) provides the most detailed source material and is analysed in depth. The researcher's access to and knowledge of this project facilitated information and data sources that are generally out of reach for most P/CVE-related interventions for reasons of attribution, security, anonymity, etc. The remaining two case studies (CC and SR) are less well documented in the public access space. Therefore, content relating to the projects has been harder to obtain. Nevertheless, sufficient information was gathered, and with the methodology framework in place, they offered exciting insight alongside AX. This analysis begins with AX and uses the criteria outlined above alongside narrative analysis, a component of qualitative research design within social science disciplines.

References

1. Bernstein, B. J., Allison, G., & Zelikow, P. (1999). Essence of decision. *Foreign Policy, 114*, 121. https://doi.org/10.2307/1149596
2. Braddock, K. (2020). *WEAPONIZED WORDS: the strategic role of persuasion in violent radicalization and counter... -radicalization.* Cambridge Univ Press.
3. Campbell, J.P., Daft, R.L., Hulin, C.L. (1982). *Generating and developing research questions.* Sage Publications. First Edition.
4. Casey, L., Cb, D. (2016). *December 2016 Dame Louise Casey DBE CB The Casey Review: a review into opportunity and integration executive summary.* [online]. Available at: https://assets.publishing.service.gov.uk/media/5a80c708ed915d74e33fc58c/The_Casey_Review_Executive_Summary.pdf.
5. Denzin, N. K., & Lincoln, Y. S. (Eds.). (2018). *The SAGE handbook of qualitative research* (5th ed.). Sage.
6. Falkheimer, J. and Heide, M. (2018). *Strategic communication: An introduction.* Routledge Taylor and Francis.
7. Francis, M. (2014). *How do we prevent radicalisation? By Matthew Francis.* [online] Theos Think Tank. Available at: https://www.theosthinktank.co.uk/comment/2014/09/26/how-do-we-prevent-radicalisation-by-matthew-francis [Accessed 20 Mar. 2022].
8. Furedi, F. (2007). *Invitation to terror.* Continuum.
9. Giddens, A. (1984). *The constitution of society: outline of the theory of structuration.* Polity Press.
10. GIFCT. (2021). *Content-sharing algorithms, processes, and positive interventions working group Part 2: Positive interventions.* [online] Available at: https://gifct.org/wp-content/uploads/2021/07/GIFCT-CAPI2-2021.pdf [Accessed 18 Dec. 2022].
11. Hassan, A. (2012). Shifting paradigms. *Arches Quarterly*, 5(9).

12. Hedayah. (2021). *Counter Narrative Library—Hedayah Website*. [online] Hedayah Website. Available at: https://hedayahcenter.org/resources/counter-narrative-library/ [Accessed 19 Mar. 2023].
13. Hedrick, T., Bickman, L., & Rog, D. (1993). Applied research design. *Sage Publications*. https://doi.org/10.4135/9781412983457
14. Horgan, J., Altier, M. B., Shortland, N., & Taylor, M. (2016). Walking away: The disengagement and de-radicalization of a violent right-wing extremist. *Behavioral Sciences of Terrorism and Political Aggression, 9*(2), 63–77. https://doi.org/10.1080/19434472.2016.1156722
15. Institute for Strategic Dialogue (2016). *The Impact of Counter-Narratives*. [online] ISD. Available at: https://www.isdglobal.org/isd-publications/the-impact-of-counter-narratives/#:~:text=ISD [Accessed 19 Nov. 2022].
16. Jackson, R. (2016). *Routledge handbook of critical terrorism studies*. Routledge.
17. Ranstorp, M. (2007). *Mapping terrorism research*. Taylor & Francis.
18. Lin, S.-Y. (2018). Interpretation and the implied author: A descriptive project. *The Southern Journal of Philosophy, 56*(1), 83–100. https://doi.org/10.1111/sjp.12269
19. Ragin, C.C., & Becker, H.S. (1993). What is a case? Edited by C. C. Ragin and H. S. Becker. (Pp. 242.) Cambridge University Press: Cambridge. 1992. *Psychological Medicine*, 23(2), pp. 534–534. https://doi.org/10.1017/s0033291700028737.
20. Saltman, E., & GIFCT. (2022). *Introducing 2022 GIFCT Working Group Outputs*. [online] Available at: https://gifct.org/wp-content/uploads/2022/07/GIFCT-22WG-ExecSummary-1.1.pdf [Accessed 21 Mar. 2023].
21. Silke, A. (2010). *The psychology of counter-terrorism*. Routledge.
22. Wilkinson, P. (2011). *Terrorism versus democracy*. Taylor & Francis.

Chapter 4
Abdullah X Project

Abstract This chapter presents an in-depth case study of the Abdullah-X (AX) project as a pioneering online counternarrative (CN) effort within the broader Preventing and Countering Violent Extremism (P/CVE) field. Emerging from grassroots beginnings and evolving into an internationally recognised pilot campaign, the AX project represents a significant juncture in the strategic communications and behaviour change landscape of early P/CVE initiatives. The chapter chronicles the project's conceptual origins, the rationale for its distinctive animated storytelling approach, and the strategic interplay of content design, theological nuance, and audience engagement. It examines how AX leveraged online advertising tools, keyword-based targeting, and platform-specific dissemination methods (YouTube, Facebook) to introduce alternative narratives at critical decision-making junctures for individuals at various points along potential radicalisation trajectories. Drawing on socialisation theory and the Reasoned Action Model (RAM) of behaviour change, the chapter situates AX's thematic and stylistic choices as an intentional re-socialisation tool—offering "mental armour" against extremist propaganda. Through narrative and discourse analysis of specific videos (e.g., "Five Considerations for a Muslim in Syria") and user journey examples, it highlights how AX content aimed to undermine extremist "calls to action," raise digital literacy and recalibrate normative beliefs about jihad and religious obligations. Ultimately, the Abdullah-X case study reveals that credible, contextually rich, and strategically disseminated counternarratives can not only prevent vulnerable individuals from embracing extremism but may also spark reconsideration in already radicalised persons.

Keywords Abdullah-X · Technology sector · Algorithms · CVE—countering violent extremism · PVE—preventing violent extremism · Campaigns · Counter extremism

M. Zamir and C. Kaunert, *Strategic Communications' Role in Counter-Terrorism*, Advanced Sciences and Technologies for Security Applications,
https://doi.org/10.1007/978-3-031-89110-6_4

Introduction

The AX project remains one of the most unique CN efforts within P/CVE due to its historical place within the field, origins, impact, and eventual disappearance (the project ended earlier than intended) from the P/CVE landscape. As a standalone exemplar of the potential within P/CVE and its associated pitfalls or hazards, AX offers much by way of conceptual and directional understanding. It can be argued that this project is both a tale of incredible achievement and a cautionary tale (lessons for P/CVE online efforts) from the perspective of impact, both upbeat and less so. This project gave rise to much CN understanding or 'good practice' and inspired the two other case studies in this section and numerous others.

The Origins of AX

The project started in 2012 to relative obscurity, an endeavour created by a small grassroots CSO in East London, UK. The CSO delivered offline interventions in a one-to-one context to 'at-risk' individuals within the UK counter-extremism field and delivered ideology awareness training to frontline staff nationally and internationally. Using learning from offline interventions and a skillset based on addressing the Islamist/Jihadi narrative, the foundations of this project offered relevance to the rapid rise of social media-based extremism, online extremist content proliferation and lack of alternatives. A significant gap in online counter-extremism existed in the Arab Spring and early Daesh era. Without any external source of funding, the CSO set about conceptualising what type of creative CN solutions could be found that could attract, engage, and retain a follower base keen on a particular style, tone, and theme.

The rationale appears to have created an authentic, recognisable, and relatable brand that could deal with specific issues relating to extremism and universal vulnerabilities that affect Muslim youth in the UK context. The make-up of the CSO suggests some individuals with proximity and understanding of extremism were involved, with some of this proximity meaning 'former' extremists were involved in the work to counter extremism at a local level. Identifying a gap in the P/CVE field and seeking to fill it required a coherent concept, creative vision, and narrative credibility. Much of AX's success can be attributed to how the project realised these needs at the right time. From a research perspective, it is crucial to identify where certain conceptual, strategic, and communicative elements were woven into the design and delivery of the products created.

Conceptual Design and Broad Aims of the AX Project

The AX project developed from within the premise of tailoring to an audience (youth) that had seamlessly embraced the online sphere of soundbite-based content in the form of video, audio and written content. During its inception phase, these elements were best represented within the online world by YouTube, Twitter (X), and Facebook. What had become apparent to the broad P/CVE fraternity during the period 2010–2012 was that extremist narratives were highly prevalent on the internet, but not always immediately visible. Alongside this realisation came one that related to the dearth of CNs in the online domain. Another critical contrast to any CN efforts to this up to this point was that extremist content online could take the shape of dense sermonised or conflict-depicting video content, long and arduous uploads of religious texts interpreted by Jihadis or doctrinal instruction (dawah) websites.

The presence of extremist ideas on Twitter (now called 'X') was also becoming a worry, yet take-downs and censorship were not everyday outcomes during this stage. It appeared that the online Jihadi narrative was becoming adept at engaging, attracting, and ultimately persuading the impressionable user to 'join the cause'. Therefore, persuasion and narrative were at the centre of the extremist pitch and needed to be at the centre of the CN offer. The AX project seemingly decided on a bold and previously uncharted model within P/CVE to attempt to affect the proliferation of extremist content online. The project used animated video content featuring cartoon characters, a slick yet grainy transition style, and one central protagonist, Abdullah-X.

The central character is depicted as a streetwise Muslim teenager with an exciting appearance and someone with a well-developed sense of 'self', identity, and place. The videos featured a 'hook' or 'Mind of a Scholar, Heart of a Warrior' tagline. Comic-style visual and text boxes accompanied scenes, with audio voiceover accompanying text boxes. At a general level, the content was designed to engage UK Muslim youth (predominantly male). It featured often outlandish examples to make a critical point in the narrative arc. One of the things that is clear about CN work in the online domain is that the finished product not only becomes a manifestation of an initial aim but also a narrative in and of itself within the online sphere, extremist sphere, and P/CVE sphere.

This is something that warrants a section devoted to its discussion later in this chapter. Firstly, in order to understand the features, aims, and outcomes of AX, a detailed examination of the narratives, assumptions, and output of the project is needed. This will incorporate discussion of the theoretical framework, strategic communications context of the narratives presented, and how the efforts of AX link or do not link to behaviour change understanding, such as the RAM approach. The core AX narrative appears to be one that seeks to undermine misconceptions of Jihad, challenge social issues that exacerbate vulnerability among the target group and 'call-out' the actions of other actors in the local–global context that add to the milieu of factors present in conflict, terrorism, and victimhood/grievance narratives. An intricate analysis of the language codes, creative process and strategic intent behind the communications aired by AX is needed and warranted. Two specific pieces of

AX content will be used as samples to gain an in-depth understanding of the project's critical characteristics throughout the case study exploration and contribution to the P/CVE CN field.

Discourse Analysis of the AX Project

In the initial stage of AX's development and dissemination, there is a strong suggestion of experimentation, design, and narrative testing. This included testing creative style, duration of videos, and narrative themes. Earliest renditions of the character and style point to a clear attempt to offer knowledge-building, resilience-building, and undermining of the Jihadi narrative as central themes. It can be argued that the provocatively titled 'Real Meaning of Jihad' video (example 1) was a very early example of an inoculation-based CN aimed at exposing audiences to the diverse processes of radicalisation from a reverse angle, i.e. a process of building resistance and awareness of the tactics often used by recruiters offline and online. Example 1 thus has clear links to an inoculation/prevention approach with two exciting features added also that were rarely seen during this period of online P/CVE. The inoculation method will be discussed in detail later, but for now, this example's other features provide the discussion with some significant contextual insight into innovative approaches to CN understanding.

Example 1 contained a narrative element that incorporated AX being contacted via private messaging online by a stranger. Such a story arc mirrored elements of real-life online radicalisation processes where individuals can be targeted and engaged with by extremists with a view to indoctrinating the individual into adopting the cause. Where this example showed innovation and creativity was that the story arc then shifts the emphasis and individual's journey to the offline domain, where again the engagement, push and pull factors associated with offline extremist recruitment are presented. In manifesting online and offline contexts, example 1 offers the audience obvious and interrelated themes that can build awareness, develop resilience, and engage attitude and behaviour-based considerations in both domains. There are clear indicators within this dual function of the narrative, focusing on 'seeking' stages and shifting attitude to informing intention, which in turn affects behaviour among the target audience. This bears the conceptual footprint of the reasoned action approach and becomes more obvious when the second unique facet is introduced.

Example 1 uses a consistent thread of theological input that directly challenges and refutes Jihadi rendering of classical Islamic texts whilst simultaneously offering the viewer core knowledge-building opportunities with which to build 'mental armour' against skewed future attempts to communicate interpretations of Quranic and Hadith sources. What makes this video stand out in this regard is the open use of critical texts that had never previously been used in CN work out of fear of audience reactions, offending sectarian sensitivities and being intentionally misconstrued by those seeking to undermine P/CVE efforts. There is also the issue of such sacred texts and meanings being administered in an animated cartoon set in contemporary East London.

The strategic intent within this video points to being based on delivering theologically-driven refutations of the Jihadi narrative, packaged in an engaging and mystery story-based animation. The narrative arc charts a day in the life of a somewhat confused AX, who goes through various precarious subliminal and face-to-face engagements before his day finally comes to a provocative yet profound conclusion. He is told he has been watched lately and that some of the things he has been doing are not beneficial for him. He is then left with a parting thought: the true gift is realisation (self-realisation). Example 1 uses elements of socialisation theory through clever use of seeking to realign normative ideas within the audience with regards to personal responsibility/obligation, responsibility towards others, understanding and respecting those in authority, risk/benefit scenarios to improve decision-making, and signposting a clear boundary between what is considered 'moderate' and what is 'extreme'.

Elements of the video use emotional, cultural, and intellectual levers to weave relatable and topical issues in and out of the main story. Within the reasoned action approach, this links to the 'background and contextual factors' component of the theory, which forms the individual contextual links to intention, attitude and behaviour issues as the model progresses through its logic flow. The video does not use the term radicalisation at any stage and links the theological texts to references for further exploration or study if desired. This appears to follow the Islamic tenant of providing one's sources if communicating doctrinal matters. The video is approximately twenty-two minutes in length, which seems to be out of sync with modern social media content guidelines on duration. It is fair to categorise example 1 as a short, animated film. The original version is a unique and standalone CN product with several interlinked facets that were innovative for the period in question (2012–2013). The video did not, however, gain any traction or attention at the time, and it seems the project and its focus underwent a period of reflection and redesign before AX was to be seen again. The same video would re-emerge a year later, re-edited, and repackaged with a more apparent purpose and strategy behind its dissemination.

Within the AX case study context, the project's historical relationship with online P/CVE has significance beyond the content itself. This is because the achievements of AX coincide with the emergence of P/CVE-orientated entities such as RAN and a more concerted general effort to offer some form of networked or co-ordinated response to the proliferation of extremist content living, amplifying, and being engaged with online. These efforts sought to harness civil society's 'latent' power alongside government agencies and the tech sector. This makes more sense when names and 'faces' are explained in the context of efforts that eventually paved the way for the current state of the P/CVE field, positive and negative, precisely the CN element. To establish itself and begin the process of developing both a network and an overarching purpose, RAN sent out a call for CN proposals via a relatively small network of practitioners and think tanks who were in the P/CVE space at the time.

This call to action asked community sector activists, creatives, and organisations to propose CN ideas that could be used in an online pilot programme to build learning and good practice and contribute to creating an evidence base of 'what works' and 'what does not'. The call to action was met by the AX project, facilitated by ISD, Google Ideas, and YouTube and managed via RAN. The age of networked P/CVE arrived with little fanfare but morphed eventually into a leviathan of global proportions. Example 1 was re-edited and became a seven-minute video, peer-reviewed by the CSO and tech and communications experts and became one of six pilot videos created using the AX branding. The character underwent more creative design work, a more strategically-orientated visual look that could stand out among the echo chambers of the internet, and with a clear modus operandi by early 2014 to seek to counter the waves of Daesh-related content, support and recruitment found online. AX was an anti-Daesh narrative tool with an expansive digital playground in which to be unleashed.

It is somewhat ironic that much of the debate and conjecture around P/CVE CNs revolves around success criteria such as vanity metrics, reach and rudimentary engagement data, such as age and gender or location. The irony is that example 2 of the AX case study reveals a set of indicators of greater significance when form and content were in scarce supply and commercial evaluation models were being applied to specialist counter-extremism efforts. Example 2 highlights decisive and defined strategic intent facets, with a clear strategy regarding what is being communicated and to whom to achieve a defined end-state objective.

Even more revealing within this example is the apparent reliance on the actual narrative component being the carrier of the message and not visual or genre-related elements. Example 2 is titled 'Five Considerations for a Muslim in Syria' and lasts under three minutes. The video is less polished in production quality compared to example 1, and uses voice narration at the expense of comic speech bubbles. There are fewer transitions in terms of scenes, with the central character (AX) seemingly offering a reportage-style commentary to the viewer, combined alongside stock footage of conflict and destruction layered and treated with visual effects to appear less intrusive. A noticeable change in this example is the look and overall persona of the central protagonist, AX.

Example 2 portrays AX as a streetwise and intelligent individual dressed in a way reminiscent of characters found in the HBO animation series 'The Boondocks, alongside elements of Japanese anime style. AX has three braids in his hair, big eyes and a chain with the logo 'C7'. Whilst none of these elements appear to have any meaning in and of their right, further analysis introduces enhanced strategic communications features and aspects of indirect 'visual-narrative' woven into the character's overall look and persona. Using semiotic analysis, it is possible to delve deeper into the representation of the AX character and assess the construction of meanings, reference points, subtle use of language, and cultural and religious cues found within his depiction. This granular analysis will also allow a more robust discussion around socialisation theory and RAM application, central to the research project's epistemological and ontological aims. A detailed exploration of the character and narrative design elements using sociological tools such as semiotics and aspects of media theory can hypothetically reveal actual tangible components of strategically orientated design and communicative processes. These would then highlight the role of strategic communications within behaviour change efforts, couched within the broad theoretical arc of socialisation processes, rather than attempting to fit P/CVE CNs into radicalisation theory-based understanding. The rationale is that radicalisation is too convoluted and confusing as a stand-alone concept to predicate such important work without resting on more established theoretical footings, such as socialisation, re-socialisation and political socialisation.

Framework and Criteria Analysis: RAM, Socialisation, and Communication Models

The chosen methodology for this research project focuses on qualitative discourse and narrative analysis via case study exploration. This is achieved using a theory framework incorporating the RAM, socialisation theory and Communications studies into the SoRaSCo model. The rationale for choosing this approach stems from a need to understand CNs from interpretive, strategic communications, and behaviour change perspectives. With too much reliance on metric evaluations and assessments of P/CVE CNs, this chosen methodology seeks to understand what makes CNs effective beyond likes, shares and reach. What components of strategy, design, narrative construction (where applicable) and communications delivery can be identified to build out a conceptual framework for future efforts in this field? At the juncture of detailing example 2 of the AX case study, this requirement is more prominent, given the strategic relationship between the AX project, P/CVE efforts during the development of RAN and the rise of the tech sector involvement in addressing security needs on their platforms. It can be proposed that example 2 of AX, connected to the abovementioned processes, will be seen as an effective CN product and evidence of 'good' P/CVE or ineffective. To arrive at a tangible conclusion in this regard, analysis of qualitative and some quantitative data (where available) will be required.

The first matter to analyse is the character's name and project, Abdullah-X. Extremist narratives that base themselves apparently on 'Islamic' sources tend to stress a message of pure, unfiltered obedience to Allah. The narrative's core emphasises that authority is for God/Allah alone (Hakimiyyah) and that the 'righteous' believer or adherent will reject 'man-made' systems such as democracy or the rule of law where such governance is in place. The narrative fails to mention, however, that the adherent does not have much say on who they should take orders from on behalf of Allah. In essence, authority is for Allah alone. Still, the follower needs to give allegiance and loyalty to a man in the form of an 'Amir' (leader), who has seemingly interpreted the wishes of Allah to mean indiscriminate killing and rejection of anyone or anything that does not comply with this specific 'doctrinal' position. This is then enforced through the process of declaring non-followers as apostates (murtad), worshipping or following other than Allah, which means being labelled as 'taghut', such as elected leaders or institutions, which then enables a judgement of excommunication (takfir) upon them. At this point, these non-followers are no longer protected as 'true' Muslims and, therefore, become legitimate targets for the group. Why is this context relevant to the name of the character and project?

Abdullah is translated to mean 'servant of Allah'. It is a revered name within the Islamic lexicon, recognisable and culturally broad in its uses beyond race, ethnicity, and language variation. The interesting point of note here is that this name carries significance symbolically for being associated with servanthood and its relationship with divinity. It is thus a name with positive meanings from a historical and linguistic position, meaning precisely what its translation implies. The 'X' moniker carries two interlinked meanings, both associated with narratives linked to struggle, activism and work towards change. Firstly, from the content and its themes, AX appears to be someone who reveres and respects the legacy of Malcolm X as a role-model type figure or inspiration. Malcolm X was clear that the X represented 'no name' due to the rejection of adopting or taking a slave owner's name. In this regard, X can come to symbolise 'no name' or 'unknown'. Taking the latter variant, Abdullah-X is the servant of Allah, unknown, or in adjusted technical grammatic translation, the 'unknown servant of Allah'.

This could mean that AX is an unknown person or can also be interpreted to mean anyone could be AX, if they are motivated to learn and understand the reality behind the issues blighting Muslims. In this way, the naming of AX offers elements of universal appeal and specific appeal to those who may feel disempowered or voiceless in the face of everyday struggle or more profound issues such as Islamophobia, racism, and conflict. This sets up a more defined understanding of why the narrative of AX comes across a certain way. AX gives away the rest of this understanding without much need for deeper analysis through the tagline; 'mind of a scholar, heart of a warrior'. The tagline does, however, warrant further elaboration in the context of what the narratives presented by AX signpost viewers to in terms of ideas.

Kurt Braddock's analysis of AX points to a character who is comfortable in a Western-centric Muslim context with issues around identity management, a sense of belonging, and what loyalty and duty may look like. AX has no problem with expressing what it means to be a Muslim living in the West and does not mind calling out contradictions, double standards, or behaviours that go against his moral code [5]. Although provocative at first glance, the tagline seems to morph two elements of theological narrative, military lexicon, and 'grand strategy' thinking, as championed in the West by the Clausewitz school of thought [13] and in Eastern tradition by Sun Tzu [19]. There is a suggestion that intellectual prowess is necessary for a young Muslim to develop in theological, legal, and worldly knowledge banks. This echoes the famous hadith of the Prophet Muhammad, "seek knowledge as far as China" [2], in that it is a Muslim's personal responsibility to seek appropriate forms of knowledge and develop these sources through the cultivation of his/her intellect or 'aql' in the Islamic tradition. There is an implied emphasis on gaining a wider berth of knowledge than just relying on what someone is told by another or through one information source. This links to the idea of plurality in thought acquired through knowledge and applied through reason and intellect. AX is himself, after all, the unknown servant of Allah, so he is not telling anyone to 'follow him' either. The latter part of the tagline appears to be linked to the difference between a 'soldier' and a 'warrior', which in a linguistic sense, is an important contrast given how Jihadi networks champion the role of the soldier, but downplay the issue of a warrior, as the latter can mean differing levels of rank, authority and intellectual gravitas.

In military discourse, a warrior can be a general, captain, or infantryman/woman. The term relates more to characteristics and traits as opposed to roles and expectations. In sum, then, the tagline seems to champion the role of knowledge through intellectual strength as the key to 'battling' through struggles, disinformation, and distortions of faith and everyday life. This leads the symbolic analysis to other visual elements of the character's look and appearance in anticipation of interpreting example 2 from a narrative angle.

As has been suggested, AX could technically be anyone, so what significance is there to the design makeover the character underwent at the start of the RAN/YouTube pilot campaign in 2014. Two interesting issues emerge from relatively hidden or more obscure elements of the character design. The first is the hairstyle adopted by the 2014 variant of AX. Sporting three equally placed braids, AX exudes a street culture-type persona that can gravitate towards young, impressionable, and fashion-conscious audiences. The character looks bold and larger than life, something an animated cartoon has to be if it is to pierce through existing stereotypes and well-developed socialisation processes of conformity, social norms and values. Equally, however, this hairstyle has another more profound symbolism behind it. There are historical references in Hadith literature that the Prophet Muhammad wore his hair in braids (three) during specific battles. AX's hairstyle and beard suggest a character steeped in tradition yet comfortable in the modern. In fact, it is fair to suggest the character possesses both the uniqueness of individuality and the awareness of commonality, which makes him appear to be 'above average' in thought and deed. AX's references are explicit yet often subliminal, suggesting a well-developed sense of self, identity and place.

This also applies to the name the CSO uses in its marketing: Cypher 7AD or C7 in abbreviation. Cypher is a term also used to denote a circle or type of code. The 7AD element denotes the advent of Islam in the seventh century AD. This means the brand name behind AX is a code or circle representing the birth of Islam, upon which the first adherents would meet in small private circles (halaqa) to be given religious instruction (dawah). This symbolic reference is more relevant to the research project in the critical analysis section of the offline/online nexus of P/CVE interventions, as well as the CC pilot case study section. A complete and detailed discussion of theological inputs within intervention methods will unpick the extent to which P/CVE and CNs have embraced or looked away from using theology as a critical tool in safeguarding vulnerable individuals. What is clear now is that AX's design and references were clearly and strategically thought out to be eye-catching, curiosity-raising, and engaging for audiences. Behaviour modelling and reasoned action elements are visible within this context, but these will be addressed further at the end of this case study alongside links to socialisation theory. The following discussion concerns the narrative of Example 2, 'Five Considerations for a Muslim on Syria', the video that changed the P/CVE CN landscape.

Pressure was mounting on governments within the EU to seek to wrestle and contain the high numbers of people leaving Europe to join Daesh and address the enormous volume of pro-Daesh content found online with specific reference to social media channels. Regardless of the apparent dangers of private end-to-end encrypted messages services being exploited by the network, the issue of publicly available content had caused concern and fear at the cumulative impact of this content on vulnerable audiences and the recruitment numbers of 'foreign fighters' (the term given to people who left their homelands to join Daesh). Left unchecked and unchallenged, the argument was that Daesh would have almost free reign to spread, reinforce and recruit large numbers of people to its cause solely by using the power of social media [14]. From within this milieu of political, policy, and public anxiety, the AX project offered the problem set a communications-based attempt at challenging the core narrative arcs espoused by Daesh within the online space. To summarise, these were: victory is promised for the true believers, it is a duty to fight for and maintain the Islamic State (Caliphate), and Muslims will be able to live under a pure and protected land, free of man-made rule and authority [14]. Much of this central narrative was packaged with modern language and popular culture references that would be relatively easy for specific youth audiences to recognise, engage with and share. Within the core narrative presented, no mention of the geo-political reality in the region took place, the in-fighting and racial discrimination among the group or the treatment of other Muslims and non-Muslims. Alongside this denial, day-to-day living and administrative issues were swept under the carpet. This was a full-on propaganda-based communications effort on the part of Daesh and its affiliate media.

The consideration for the AX video in question was how to offer some form of undermining of this narrative, challenge the presumptions of those leaning towards it, and provide an alternative viewpoint within a short-animated video from a character no one knew or had heard of. From a theoretical framework context, this meant seeking behaviour change through offering rational and relevant strategic communications via the RAM while using socialisation and re-socialisation concepts to support this potential suite of effects/outcomes. As will be explained later in this section, there is a distinct value and benefit in elucidating, in as much detail as possible, the facets, nuances and elements of this particular video for the purposes of outlining practices and processes that can further the understanding of the CN field. As such, example 2 will be given a complete narrative analysis using P/CVE intervention methodology, semiotics and interpretivist theory elements. One of this research project's main emergent findings is that attention to vanity metrics distracts from narrative and content quality issues within CNs. Hence, for this project and its aims, it is essential to delve into the construction and contextual relevance of CNs. Since this specific video is the most 'famous', this is an ideal sample to commence this process. A full transcript of the narrative within the video is provided below, after which a critical discourse analysis of its form and content will be conducted.

Transcript of Abdullah-X's 'Five Considerations for a Muslim on Syria' Video

"Salaam, thinking about Syria and what's going on there has become almost a necessity now, but equally thinking about the best way to support the people of Syria should be at the forefront of that process. Our young people are often being driven to act without being driven to think in detail first, so let's change this dynamic for a better outcome Insha'Allah. There are five considerations or questions that an individual could do with asking him or herself on the Syria issue.

Number one; have you thought about the actual needs of the people especially women and children in Syria? Number two; if you have, then do you genuinely think that going out there to fight and potentially be killed will affect those suffering in a positive way? Number three; do the people of Syria really need you to go out there and fight when you don't fully know the terrain, the political reality, and the actual motivations of groups who are claiming to be fighting for Islam? Number four; if you have families and even basic responsibilities where you live, how would you justify leaving them for a conflict in which Muslims are essentially killing other Muslims indiscriminately and without regard for the Islamic principles of warfare and the protection of civilian life? Number five; have you asked yourself sincerely what your actual maqasid (objective) is and if this is actually anything to do with Islam or protecting other people?

If you have, then have you not found more constructive ways to support the innocent people of Syria, compared to trying to go out there and fight, to simulate some video game that you feel you have to re-enact in real life? There is a 'call of duty' for Syria and it is to be well-informed and not misinformed. Allah does not need your so-called 'martyrdom' when the fight out there is about power and influence, in the form of some manufactured Jihad. Do you even know how fard ayn (personal obligation) and fard kifayah (communal obligation) apply to you in your homeland and to your own responsibilities? Engage in relief work for those people and raise awareness. Don't go and fight for a so-called cause you have not even questioned critically. More next week Insha'Allah on another topic that is current, or that will not go away. Abdullah-X. Peace." [1].

RAM in Conjunction with Strategic Communications Analysis

The overall objective of the video (end-state objective) appears to be 'to reduce support for Daesh within the UK among Muslim youth population'. This end-state objective can be understood from the narrative content of the video itself through a process of deductive reasoning, which is premised on the following propositions:

Daesh narrative is populating social media and the Internet in general.
This narrative is relatively simple in construct and tone on purpose.
The content attracts people through engagement and sentiment.

The level of engagement is linked to defined 'calls to action' within the narrative. The calls to action (CTA) are clear, concise and stratified (travel, support, promote).

In order to lessen the effect of the CTA, it is necessary to reduce the group's degree of support. To reduce the degree of support, it is required to weaken the group's narrative. A strategic framework containing the communicative, affective, and effect-based acts that will produce the end-state aim is required for this process to be realisable. This type of communications-based objective necessitates an effective strategy. In theory, the employment of strategic communications tools in P/CVE CN initiatives facilitates a 'positive impact' once the communicative form is in the operating environment and reaches the intended audience. By combining the strategic framework and TRA, an evaluation of Example 2's impact can not only shed light on the compelling aspects of this particular product but also serve as a foundation for future CN design requirements. In this case study, a possible strategic framework could look as follows.

Initially, an analysis of the adversary (Daesh) and its capabilities in the domain (online) would be required. This would outline intent and enable a strategy for effective counter-communications/narratives to be developed. Figure 4.1 below illustrates a possible Adversary Analysis of this kind, which is built from influence operations doctrine and strategic communications theory.

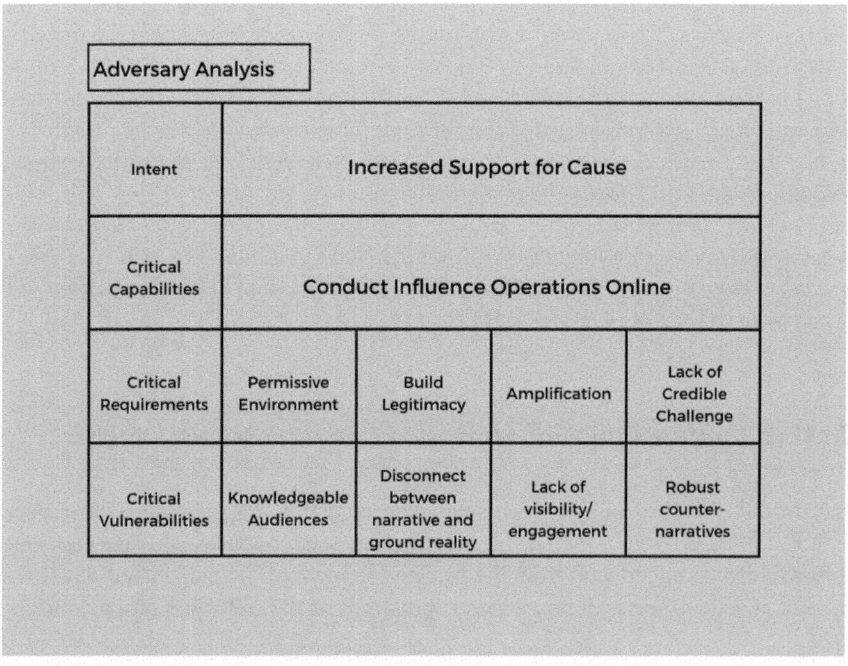

Fig. 4.1 Adversary analysis (Strategic Framework)

The need for an adversary analysis is twofold; firstly, it outlines the known intent and critical capabilities of said adversary within a particular environment (in this case, online). From a P/CVE perspective, this intent would be known from obvious or tangible actions, such as a proliferation of pro-Daesh content circulating social media and the internet in general in a 'public space' (social settings). In other contexts, such as law enforcement or military operations, intelligence gathering and covert human intelligence sources or 'stringer' networks (on-the-ground contacts) would be part of the formation of information gathering and subsequent assessment. Identifying the intent (the 'what') to a greater extent also explains the action's purpose (aim/objective), such as 'promoting its ideology', 'recruiting followers', and 'creating fear and unrest'. Alongside this, the need to understand and outline the adversary's specific capability to pursue this intent is essential. In this case, the ability to create, disseminate and engage with audiences through communications has been identified. This means the adversary can technically conduct influence operations in the online domain. However, intent and capability alone do not cover the scope and needs of the adversary or those seeking to challenge this adversary. Possessing capability without specific fundamental requirements being met or being able to exploit weaknesses would mean the context remains hypothetical and not operational. A second level of analysis is thus required to complete a relevant and practical scoping.

Secondly, therefore, understanding which critical requirements the adversary will seek to rely on needs to be identified. These requirements link to external factors the adversary can harness and exploit for its aims. Requirements are only technically met fully if no identified vulnerabilities or weaknesses can be used to reduce, undermine, or degrade the adversary's aims. As such, critical requirements cannot exist independently of essential vulnerabilities, even at the level of logic. This is why strategic frameworks are also referred to as logic frameworks in other areas of communication or influence/persuasion efforts. In the communications environment, very few elements are under the precise control of any actor, simply due to the nature of the communication-exchange-interpretation relationship.

There are too many extraneous variables in such a context, which means a requirement for one actor can be undermined by another actor simply by identifying a vulnerability within the requirement and transforming it (through communications) into a tool to be used against the adversary's intent. In the case of AX's example 2 video, 'Five considerations for a Muslim in Syria', what vulnerabilities does the video's narrative expose to impact the adversary's requirements, which in turn can negatively affect the adversary's capability to conduct its influence operations? It is hoped that such an analysis can reveal precisely what is strategic about strategic communications and how P/CVE CNs can benefit from such an approach. How does AX use strategic communications to challenge and counter Daesh's narrative? A deeper analysis is needed to understand these issues in a greater context and with relevant nuance by breaking down the narrative and its qualities.

Analysis of AX Example 2 Video's Narrative

> Salaam, thinking about Syria and what's going on there has become almost a necessity now, but equally thinking about the best way to support the people of Syria should be at the forefront of that process

[The first sentence implies this is a Muslim-to-Muslim conversation by using the term 'salaam'. AX also suggests that the issues in Syria are matters that warrant obligatory attention from Muslims through use of the term 'necessity'. Daesh narrative espouses that it is a Muslims personal obligation (fard ayn in Arabic) to fight by means of physical offensive Jihad, when in fact traditional understandings of this issue do not link physical offensive Jihad to the personal obligations of civilians. This 'personal obligation' would fall to an authorised, paid and professional military entity, such as a nation's army. From a strategic perspective, AX appears to be challenging the Daesh 'version' of personal obligation, not by dismissing the notion of obligation, but by changing its emphasis from personal obligation to communal obligation (fard kifayah), which is exactly what the issues in Syria would be for Muslims living outside of the country. These issues would necessitate care, attention, and some form of communal response through appropriate and legal means. AX also refers to the situation by using the term 'what's going on there', which is a rudimentary way of suggesting the issue is confusing, convoluted, and, essentially, a mess.

This could mean gaining a factual, truthful, and reliable account of the cumulative factors affecting Syria and its people is not realistic, so individuals need to understand this point firstly. He then brings the narrative back to the context by stating people need to find the 'best' way to support the matter and those caught up in it, not the quickest, cheapest, most aggressive or hateful, etc. *The character is already from line one suggesting people think carefully, not carelessly or carefreely. AX sets a tone and context that care and attention are needed, both in regards to the situation and to possible responses. This appears to set a critical thinking element to the matter and sets up the key messages that are to follow in the video.]*

> Our young people are often being driven to act without being driven to think in detail first, so let's change this dynamic for a better outcome Insha'Allah. There are five considerations or questions that an individual could do with asking him or herself on the Syria issue.

[AX hones in on young people, targeting the group demographic at greatest risk from exploitation by Daesh in a gender-neutral manner. By using the term 'our', he also continues to communicate to the audience from a position of commonality in faith, concern, and care. The term 'driven' implies possible links to indoctrination, brainwashing, grooming, etc.*of young people, with an emphasis on the fact that 'thought' alone is insufficient, and needs to be detailed/critical/conscious thought. By stating that this dynamic needs to change, he again reinforces a position, which in turn helps to keep the viewers' attention on the matter at hand, i.e. that young people are at risk and are being exploited. By using the term Insha'Allah, AX is placing the success of the intention and desire for change with the will of God. This confirms*

a faith-based stance as a core feature of the narrative at the outset. In a subtle yet seamless transition, he then uses modern numerical or alphanumeric examples to plant a number (five) and a set of processes (considerations) associated with each number in the viewer's mind. This tactic is used by marketers, bloggers, teachers, and businesses to help facilitate understanding, retention, and recall among audiences. By stating the intent, he wants to keep the audience engaged and wants to retain their attention long enough to reveal the five points. He then relates the five points to the issue of Syria and this time specifically mentions men and women as his target audience.]

> Number one; have you thought about the actual needs of the people, especially women and children in Syria?

[Consideration one places a humanitarian emphasis on the situation, which links the issues of Syria to the suffering, displacement, and distress being experienced by the people. It uses a reflective tone as well as making an appeal to the same faculty within the viewer. Use of the term 'actual' suggests the true nature of the conflict is based on the impact being felt by the average person. In this example they are identified as women and children, both of whom are considered vulnerable groups in such contexts.]

> Number two; if you have, then do you genuinely think that going out there to fight and potentially be killed will affect those suffering in a positive way?

[Having pointed out the plight of women and children, the narrative immediately re-engages at the individual level thought processes regarding potential options for action. There is an interesting interplay in this moment between the character's assessment of who may be watching the video and the issue of 'foreign fighter recruitment'. Consideration two appears to be making the link between understanding the actual situation in Syria, and conclusions being reached that leaving one's homeland to go and 'fight' (join Daesh) as being a common outcome of engagement on the issues. Put another way, AX is seemingly speaking to people who have thought about joining Daesh or are currently thinking about joining. In the latter scenario, he seems to be indirectly stating that such people are engaging in viewing pro-Daesh content online, which could distort the truth and indoctrinate them into the ideology and cause. AX is thus appealing to pre-existing socialisation processes within the audience in an attempt to thwart re-socialisation processes occurring through the consumption of terrorist content on the internet. Use of the term 'genuinely' prior to use of the term 'think' is immediately requesting the viewer to 'think again'. Just in case the point is missed, AX mentions 'going out there to fight', which clearly implies leaving another land to travel to Syria, and then he asks a very critical question for the viewer to consider; the viewer is asked if joining Daesh, fighting for its 'cause', and potentially being killed will affect the people in need in a useful manner. AX is asking viewers to question their motives.]

> Number three; do the people of Syria really need you to go out there and fight when you don't fully know the terrain, the political reality, and the actual motivations of groups who are claiming to be fighting for Islam?

[Consideration three moves the previous point forward by emphasising the geo-political situation in Syria being one of confusion and subterfuge. He then uses the term 'motivations' to question the intent of groups (Daesh) and links this to religious narratives used by such groups. There seems to be a subtle suggestion that claiming to be representative of Islam does not mean the claimant is. This calls into the question the idea of where such legitimacy comes from and how it is arrived at. This in turn, is asking the viewer to reflect on the process and challenge his/her own thinking on such claims.]

> Number four; if you have families and even basic responsibilities where you live, how would you justify leaving them for a conflict in which Muslims are essentially killing other Muslims indiscriminately and without regard for the Islamic principles of warfare and the protection of civilian life?

[Consideration four challenges the motivations of people thinking about joining Daesh by calling into question the activities of the group in regards to violence, the idea of what personal obligation (fard ayn) means in mainstream/traditional understanding and makes a key point about technical/legal aspects of warfare within Islam. The rationale here seems to be linked to making the viewer think about such actions in the context of having the appropriate knowledge, awareness and understanding on complex matters of jurisprudence regarding the Islamic concept of Jihad, warfare and the sanctity of life. AX is suggesting such matters are not arbitrary, but, on the contrary, are subject to significant constraints of legal ruling, precedent, and context. Killing civilians is not part and parcel of this legal context, and, therefore, the actions of Daesh are in direct contradiction with Islam's principles of warfare.]

> Number five; have you asked yourself sincerely what your actual maqasid (aim/objective) is and if this is actually anything to do with Islam or protecting other people?

[In the last of the five considerations, AX reaffirms that individuals need to question and reflect on not just their motivations, but equally on their respective objectives—what they hope they will achieve by their actions. He then makes a critical intervention in the process by opening up individual motivations, intentions, and behaviours as being the key to where people find themselves practically and situationally on the matter as a whole. This set of points appears to be setting the viewer up for a final push in terms of perspective, perception and position on these matters. As will be discussed in the post-narrative analysis segment, this set of considerations bear the hallmark of the TRA.]

> If you have, then have you not found more constructive ways to support the innocent people of Syria, compared to trying to go out there and fight, to simulate some video game that you feel you have to re-enact in real life? There is a 'call of duty' for Syria and it is to be well-informed and not misinformed. Allah does not need your so-called 'martyrdom' when the fight out there is about power and influence, in the form of some manufactured Jihad. Do you even know how fard ayn (personal obligation) and fard kifayah (communal obligation) apply to you in your homeland and to your own responsibilities? Engage in relief work for those people and raise awareness. Don't go and fight for a so-called cause you have not even questioned critically. More next week Insha'Allah on another topic that is current, or that will not go away. Abdullah-X. Peace." [1].

[The concluding part of the narrative is delivered as a searing portrayal of not only the motivations of people who join such groups but also the attempts by such groups to tap into and use popular culture references to engage people. This approach of using the zeitgeist has been a vital tool for Daesh in terms of getting people to support its cause in diverse ways, and any narrative seeking to counter this approach would need to address or challenge such assertions. AX seemingly does both in that he challenges the 'gamification' of the Jihad narrative as well as layering these severe issues with core theological language, which implies the need for knowledge and exploration before seeking to reach conclusions. This section of the video uses the adversary's narrative to unpick contradictions, anomalies, and assertions to demonstrate the type of skillset and approach someone would need to be well informed on issues like Syria. AX references the 'Call of Duty' video game narrative used by Daesh to engage young, impressionable men and then changes its focus to telling viewers their real responsibility is to be well informed. He challenges to use of the term 'martyrdom' and calls into question the idea of this being an act that would please God in such duplicitous circumstances. He references personal and communal obligations as matters needing to be understood and applied by the individual in his/her homeland first. This links such matters to citizenship and social responsibility. These are then brought together with a suggestion that raising awareness of the plight of the people of Syria and offering humanitarian aid is a positive way to help them. Such a suggestion serves as a call to action and brings the video full circle to its starting point again. The video ends with a 'hook' that AX will be back. This offers the viewer basic levels of motivation and intrigue, which can help to satiate the appetite for consuming video content on this matter and related issues or possibly drive the viewer to their respective knowledge and awareness of the problems.]

Having analysed the narrative AXE presented regarding its references, assertions, and main arguments, the relationship between how a CN effort such as this links to compelling strategic framework design needs is required. This process of identifying how the strategy should lie at the centre of CN design will assist in understanding the potential impact of a video such as example 2. The issues of impact in regard to this example will be discussed once the strategic considerations have been outlined. Following the outlining of the adversary analysis in Fig. 4.1 earlier in this section, the next logical strategic task would be to plan and identify the elements of communicative action required to be enacted to achieve a desired end-state objective. As such, the next stage of strategic communications planning includes identifying an achievable end-state, which also considers what certain decisive conditions would need to be met to realise this. Figure 4.2 captures these needs alongside the effects that the communications effort in question could create.

In this example, the desired end-state objective is identified as 'support for Daesh is diminished'. The most logical reason for why this is the end-state would be because, within the context of online P/CVE, fewer people watching and engaging (liking, sharing etc.) with pro-Daesh content would, at a strategic level, correlate to fewer people 'supporting' the group. As this is a strategic communications campaign, the overall criteria for success need to be linked to communications objectives. A critical potential objective in this example would be fewer people engaging in pro-Daesh

Effects Framework

End State	Support for Daesh is Diminished			
Decisive Conditions	Permissiveness of online space is reduced	Rejection of Daesh Narrative		
Objectives	Increase audience resilience levels	Build awareness levels of audience	Expose flaws in narrative and reality	Degrade value proposition
Effect	More Knowledgable Audience	Disconnect between narrative and ground reality	Lack of visibility/ engagement	Robust counter-narratives are preferred
Sub-Effect	Promote pluralism and critical thinking	Amplify weaknesses in narrative	Promote counter narratives	Recognize core ideas found in narrative

Fig. 4.2 Effects framework (for counter narrative product—Example 2)

content, which could mean less support for Daesh, impacting the Daesh brand value online. For this objective to be met, two decisive conditions are identified as needing to have manifested or to be in place. In this case, having a permissive environment for Daesh's narrative to flourish gives the adversary a significant advantage. Equally, if online communities do not actively reject Daesh and its narrative, the group has even further space to grow and evolve online and to attract attention and support. Therefore, if and when the online environment reacts against the pro-Daesh narrative by challenging, rejecting, and possibly even flagging it, this would demonstrate that counter-Daesh narratives perform positively online. A less permissive environment and greater rejection of the Daesh narrative would point to the end-state objective being met. The decisive conditions and end-state objectives require identified and measurable objectives to be worked towards. These objectives are derived from different elements of what communications-based efforts can pursue.

They are often few and made up of audience-based and adversary-related objectives so that both depth and context are covered in terms of communication needs. This small number also offers a logical approach to achievable, measurable effects, given the need for control over the different types of content and dissemination (platform) choices. In addition, working towards meeting stated objectives offers campaigns measurable and meaningful targets when brought together under specific communications products (messaging). This overall objective-led system can build a picture of how a message has performed overall but also capture specifics. In the current framework example, four key objectives are identified as central to achieving two decisive conditions, which in turn provide a guide to whether the end-state objective has been met. Each of the four objectives has specific needs to be met within this overall effort. The first is linked to increasing resilience levels among the target audience.

The logical flow and strategic benefit of this are built on the premise that the higher the resilience, the less vulnerable the audience will be to Daesh messaging. In this manner, higher resilience is a risk mitigation objective within the framework. The second objective seeks to raise the overall level of awareness of the audience through messaging types that are informative, factual, and myth-busting-based. Thus, the audience needs to be better informed of matters. The third objective is related to messaging processes that expose and, therefore, undermine the adversary's narrative. The fourth objective links to the direct challenging of the proposition or 'promises' made in the adversary's narrative to lessen the overall value of the message. The strategic use of objectives helps to outline what types of effect or output will be created when delivering against each of them. In strategic communications and persuasion-based contexts, effects and sub-effects need to be captured, correlated, and analysed so that indicative 'measures of effect' are available against 'measures of performance'. Put another way, which behaviours changed concerning how the messaging performed over time?

More resilience may be created in terms of the objectives due to the audience becoming more knowledgeable about overall matters and themselves. The audience may manifest this increased knowledge through engagement with ideas around plurality and critical thinking, which would be sub-level or secondary-level effects. Similarly, greater awareness of the situation may manifest in audiences being able to decipher the disconnect between the adversary narrative and the reality on the ground. This may be evidenced in the audience exposing flaws and weaknesses in the adversary narrative or using CN content to do so. This could also lead to effects whereby the visibility of adversary narratives is lessened, and CNs become more prevalent content streams for the audience to share, engage with, or access. Another effect would be that CNs become the narrative of choice and are even used to challenge and undermine the adversary's claims directly. This way, a block of effects and sub-effects are identified as measures of effect and monitored accordingly within the dissemination environments (platforms). A comprehensive effects framework is thus created and measured against, as is demonstrated below, with primary effects and supporting effects in tandem (Fig. 4.3).

Having applied a narrative analysis and strategic framework to the AX product and outputs, the overall discussion of the case study can take on a new stage of enquiry. As previously highlighted and a core thread throughout this research project, strategic communications are generally required to help impact some form of behaviour change among a target group or audience. The link between the communication product or message and behaviour change is two-fold; firstly, the message needs to be received by the audience with an understanding of what the critical components of the message are. Put another way, does the audience know what they are watching? Secondly, what do the message creators want the audience to do once they have

Supporting Effects

Effect	More Knowledgable Audience	Disconnect between narrative and reality	Lack of visibility/ engagement	Counter-narratives are preferred
Sub Effect	Promote pluralism and critical thinking	Amplify weaknesses in narrative	Promote counter narratives	Challenge core ideas in Daesh narrative
Supporting Effects	Respect for other views	Raised awareness levels of human rights issues	Increased viewing of alternative content	Change in perception of Daesh
	Increased ability to verify sources	Ability to identify double standards	Reduction in engagement with Daesh content	More aware of alternative ideas and views
	Enhance awareness of critical thinking skills	Able to unpick narrative/find weaknesses	Lack of interest in extremist views	Able to challenge/ refute Daesh narrative

Fig. 4.3 Sub-effects and supporting effects

received and understood the message and placed it within their cognitive space? The answer to the second part of this is that the message ideally needs to move from the mental space to the behavioural and actualised space. But this means the message should elicit a desired behaviour change. To do this, the message needs to apply a level of influence and a level of persuasion inappropriate measures by affecting human traits such as motivation, intent, and attitude.

For the strategic communications-centred fields of work within P/CVE and especially within the CN context, the RAM has been used to design and deliver communications because this approach seeks to affect the intentions by which people consider, approach, and then action their behaviour through beliefs about said behaviour [5]. In essence, the RAM approach seeks to alter the behavioural intention of the individual audience member through an intricate process of affecting norms and assumptions about norms, which, when combined with intentions, create a belief towards a specific behaviour. Using AX's example 2 video, it is possible to construct a RAM analysis that captures possible pathways by which someone exposed to example 2 (five considerations) could alter their intended behaviour trajectory in favour of the AX narrative rather than the Daesh narrative. The RAM developed by Fishbein and Ajzen is best understood with access to the model in diagrammatic form in order to understand the processes and relationships between beliefs, intentions, and behaviours (Fig. 4.4).

According to its creators, the key to understanding the RAM approach is that regardless of an individual's background or contextual factors, it is the development of

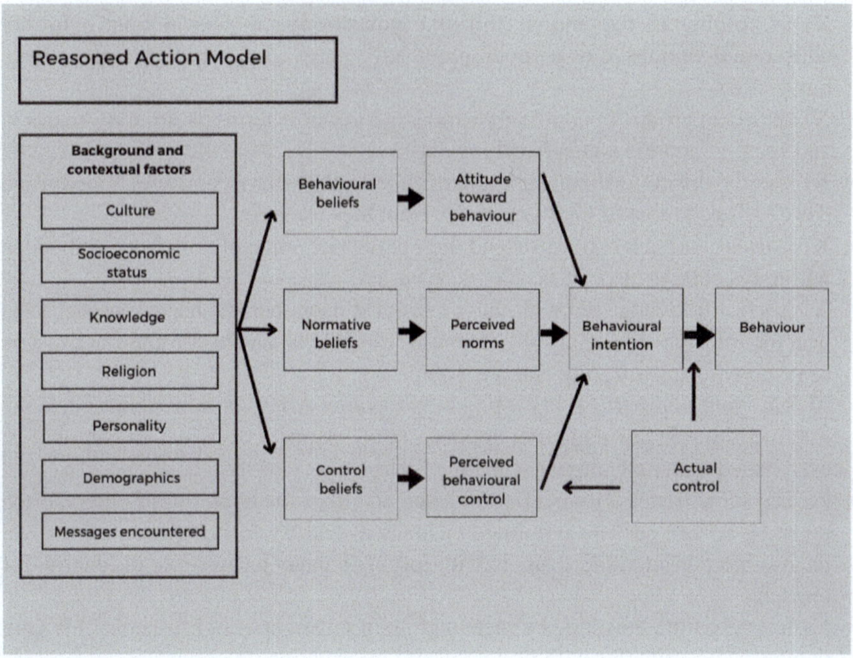

Fig. 4.4 The reasoned action model

behavioural beliefs in combination with normative assumptions about the perceived acceptability of the behaviour that needs to be addressed. They can be addressed using another layer known in the model as control beliefs and the level of perceived control the individual regards as having over these beliefs. The implication asserts that if the control beliefs are intervened on or affected, they can change behavioural intention, thus changing the behaviour. This is due to the message affecting the level of PBC the individual subsequently believes they have. From a communication perspective, this would mean the message had changed the context and intentions behind the behaviour originally being considered before the communication intervention came into the person's sphere of influence. The RAM is a systematic and process-driven approach to behaviour change that will be critically examined later in this chapter. At present, and to develop the research hypothesis further, it is essential to apply RAM to the AX video (example 2) and chart possible ways by which the messaging contained in the video could affect the behavioural intentions of someone wishing to join Daesh, through changing their attitude towards how much control they believe they have about genuinely this being doable. For this analysis, the individual will be hypothetically constructed as male, eighteen years of age and living in London, UK. This person is known as 'X1', and the following background contextual factors apply to him:

- X1 is coming to the end of full-time education and does not have further educational options due to poor predicted grades and a lack of offers from universities
- X1 has a peer group of people very similar to himself in terms of ethnicity, outlook, and socio-economic status (working class)
- X1 spend a disproportionate amount of time on the internet and social media and has developed most of his worldviews from these sources
- X1 is aware of local extremists and their networks, some of which are embedded within his college also.
- X1 has not fully engaged with these extremist peers but has been approached to become 'more aware' of his obligations towards Allah and his brothers and sisters in Islam who are suffering and need help
- X1 has been told part of this 'obligation towards Allah' is to live under Islamic rule, with Sharia law and strict application of Islamic values, which can only be established under a Caliphate (Islamic State)
- X1 has some issues with over-confidence and likes the attention of others, often meaning he can seek to dominate encounters and try to impose on others, yet he also has considerable struggle with forging a meaningful sense of identity for himself
- X1 has very little developed understanding of Islam, does not 'practise' his faith regularly and struggles with finding useful information on important issues aside from rudimentary knowledge
- X1 does like to behave and communicate in a manner that is contrary to his knowledge level and often speaks to others in an assertive and judgmental tone both online and offline, using his access to the internet at home and on his mobile phone

The first part of applying the model requires outlining specific background and contextual factors. This exercise facilitates the creation of a link between essential causal and affective relationships relevant to the individual case (X1). During this first stage of the RAM process, identifying the underlying issues that may contribute to X1 forming certain beliefs about potential behaviour choices needs to be mapped. Figure 4.5, below, demonstrates how the situational reality of X1 begins to exert influence on his state of mind regarding creating, increasing, or pushing him towards developing ideas about his beliefs and how these may need to be underpinned by certain behaviours. At this stage, X1 has yet to form a concrete understanding about either his beliefs toward future behaviour or what elements of this behaviour may even manifest. What he is doing, however, is starting to piece together aspects of motivation, intention, and situation into a loose connection of developing beliefs that are shifting his propensity and openness to explore specific paths further. When this loose connection of beliefs forms into an overall attitude about a possible type of behaviour X1's propensity to engage in this behaviour can logically start to rise within the RAM approach.

It is clear from the process outlined above in Fig. 4.5 that X1 has started to develop both behavioural beliefs and an overall attitude towards behaviour. He is basing this

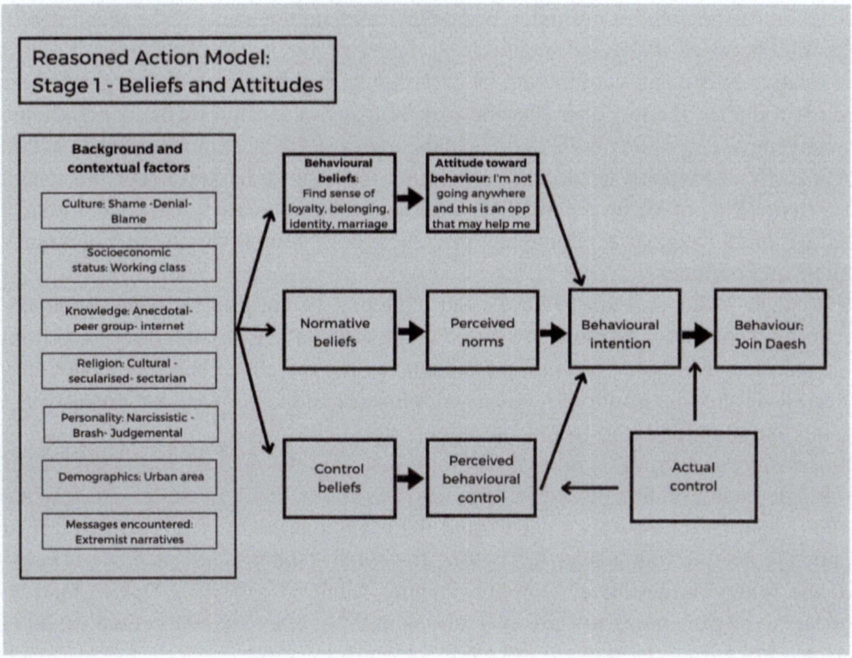

Fig. 4.5 RAM—X1's contextual factors, attitudes and behavioural beliefs

attitude on the assumption that he needs to find a greater sense of identity, which may lead to a stronger sense of belonging and possibly allow him (in his mind) to exercise a sense of loyalty and direction. The issues of an individual seeking out identity, belonging, and loyalty/duty narratives in the context of extremism and recruitment into terrorist networks remain meta-narratives that are under-developed with terrorism research but are essential elements of why people join groups, develop extreme views, and are willing to engage in actions that can include violence. Identity, belonging, and loyalty-based issues can create, drive, and bind individuals to intentions and motivations towards pursuing these needs. When personality dispositions are added to this process, e.g., someone seeking adventure, thrill, revenge, or status, attitudes about what behaviour will fulfil such needs form [20]. In X1's case, he is starting to form the attitude of believing that he needs to find an outlet for his needs and may be leaning toward the narratives he is encountering online regarding Daesh and the promises in the videos and content disseminated by the group. The next stage of this with the RAM approach is to outline how the individual develops beliefs about normative practices from within his social world that form via personal experience and perceptions about himself and others, alongside the values and norms he thinks apply to his situation.

RAM relies heavily on a conceptualisation of socialisation theory, given one of the model's core process phases is about the relationship between normative beliefs and perceived norms, both of which link to how socialisation processes impact individual ideas about acceptability, normalcy, and behavioural homogeneity in the social world. This link between RAM and socialisation theory requires critical analysis to assess the relative effectiveness of the model in the P/CVE context and, in doing so, offers a clear route for the theoretical framework of this research project to conclude the effectiveness of applying both theories to the use of strategic communications within P/CVE CNs. This analysis will take place after the case studies have been presented. Phase two (Fig. 4.6 below) of the RAM approach and X1's case brings the normative context into the equation, adding an intricate layer of complexity to the individual's beliefs and attitudes.

In stage 2, X1 appears to have been engaging in viewing pro-Daesh content, binary in terms of its narrative arc (good and bad/them and us) and offering actionable solutions for those consuming the content. The narrative, therefore, is not only offering X1 the opportunity to 'question' himself and his purpose, but equally, it is giving him the premise to form ideas about 'right' and 'wrong' based on a claim of 'superiority' and 'truth'. Through carefully distorted theological arguments, X1 is taking in the notion that those Muslims who answer the 'call' of truth via Daesh are 'true' Muslims because they understand their 'obligations' towards Allah but just happen to need to follow a terrorist group to actualise them. In essence, X1 is being fed the following narrative; 'you are obeying Allah by following Daesh', and he appears to be gravitating towards this rationale. What other elements could expedite this process during the stage of the model being discussed?

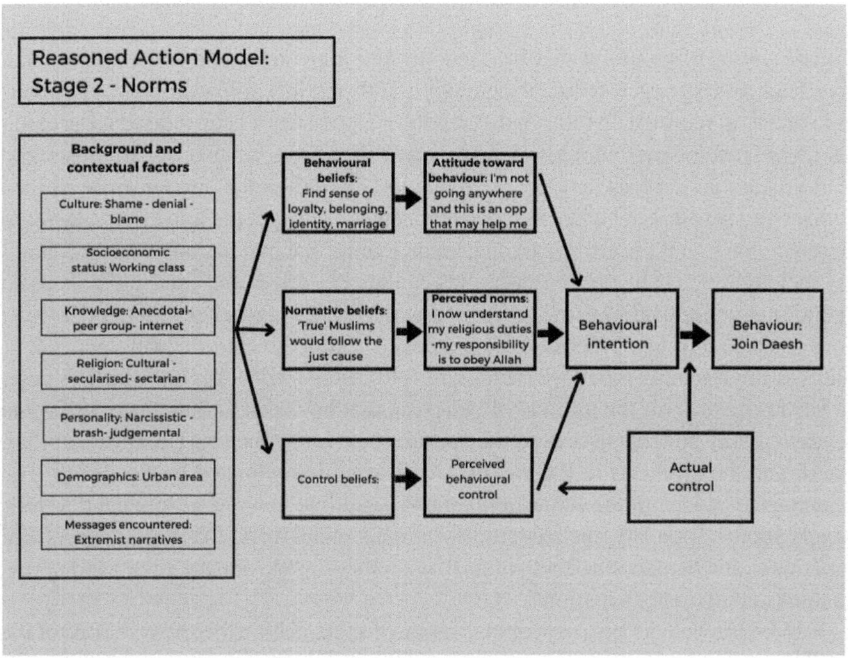

Fig. 4.6 Normative beliefs and reasoned action model

The role of socialisation processes has often been limited in academic discourses to primary and secondary variants [7]. As the study of socialisation grew within social science disciplines and took into consideration the need to understand why people join new social movements, religious cults, etc., knowledge regarding the transmission of norms and values, creation of social roles, social processes, and hierarchical relational dynamics, the theory of socialisation expanded to include a key consideration; socialisation takes on a lifelong manifestation through different and specific features of norm acquisition and values alignment [6]. People come into contact with, experience, and perceive normative ideas in various ways. Along the ageing process, social changes affect how norms and values are understood and seen as relevant. In X1's case, his social reality includes a family dynamic that espouses quite basic conspiratorial ideas about the world, denial regarding specific geo-political issues, blaming authority and democracy for everyday problems, semi-prejudicial ideas about other races and quite isolationist and politicised renderings of Islam.

Whilst these issues are not singular to any group within society, they do often drive the very polarised narratives that eventually motivate some people to adopt hate and violence as 'solutions' [16]. If X1's family narrative was summarised, it would be a culture of blaming and othering. This dynamic links to elements of socialisation from two perspectives: Firstly, it provides a foundational basis by which X1 starts to create normative assumptions about the perceived acceptance of some of his ideas, developing attitudes and behavioural intentions within his social world. He begins to think that becoming an extremist is what others would do if they, too, were 'rightly guided'. Secondly, this socialisation process includes elements of individuation, which renders the ideas being developed as additional layers of norms and values on top of pre-existing 'mainstream' ones.

In this way, X1 is being individually re-socialised. With element one of this two-factor issue taken into account, X1 is starting to normalise extremist ideas within his mindset through interpreting his socialisation within the family and peer-group network based on his online engagements with Daesh ideology, to ultimately create two simultaneous/parallel socialisations that are each at the opposite of his mainstream socialisation acquired through school and other agencies that reinforce dominant hegemonic narratives [3, 4, 10]. X1 is now starting to believe that he does not need to gain the consent of his parents, as his duty is to Allah alone. He takes the 'mentality' (mistrusting and misinformed) of the family dynamic at face-value. Then he applies his 'pure' and 'righteous' version gained from online indoctrination, telling himself this is what all 'good' and 'true' Muslims would do. He is technically undergoing three socialisations: mainstream Western, socio-cultural, and violent extremist/terrorist. The latter seeks to supplant the previous two, which is where the third stage of RAM comes into focus in Fig. 4.7.

X1 has begun to assess precisely how and through which means he can exercise his behavioural intention, having already created a behavioural attitude and normalised his beliefs in the re-socialisation processes described in stage 2. To be able to act, his behavioural beliefs, factors or elements that have control parameters attached to them, such as having access and having the means to act, would need to be within

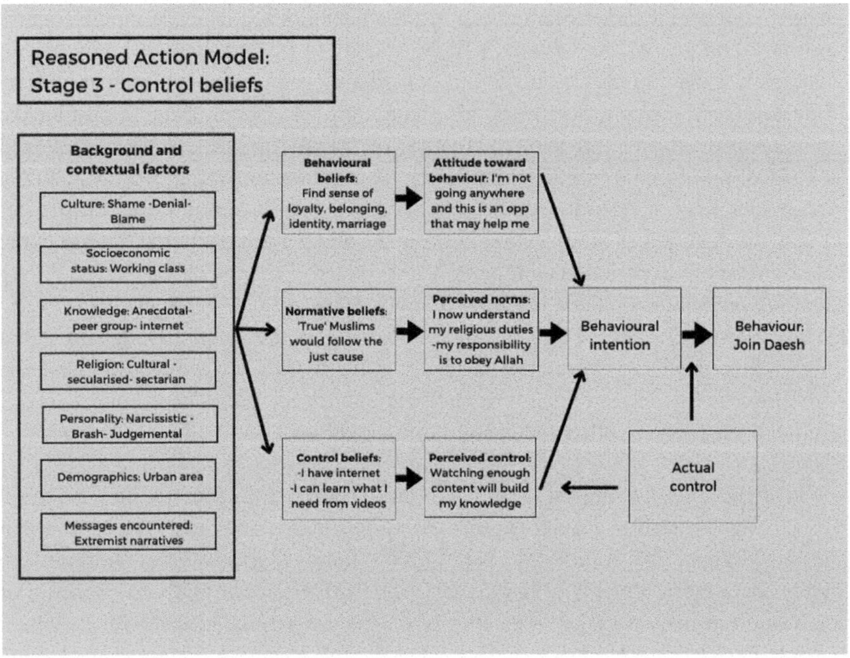

Fig. 4.7 Control beliefs and reasoned action model

his sphere of influence. The RAM approach implies that unless the individual can control the parameters (or at least exercise an element of control) that are linked to the successful pursuit and actualisation of the behavioural intention, it is difficult for the behaviour to manifest, given how strange and volatile other factors can be in this regard. In X1's case, he needs to have certain beliefs about how much control he has over both process and choice and if he can manage access to certain types of information whilst remaining under the radar to an extent within his family and friendship circle while he is gathering his knowledge and 'preparing'. These control beliefs correlate to the perceived control X1 feels over the elements needed to act on his intention of joining Daesh. If he controls the situation, he makes choices corresponding to his intentions and needs.

Thus far, he has relied on having access to the internet to engage in pro-Daesh content and to assess which online users may be like-minded based on the number of likes and follows and the types of comments they have left on videos. AX believes that watching enough content will give him the control he needs to make the choice he intends to make shortly. Some peripheral issues would also be relevant, like having money, logistics, and documents to be able to leave the UK, but X1 believes these are also under control, as he perceives likeminded individuals will assist where needed as long as he controls the process to build engagement and trust with them through watching the same content and sharing sentiments. The ultimate set of considerations in X1's mind is, therefore, whether he has seen enough content and engaged enough

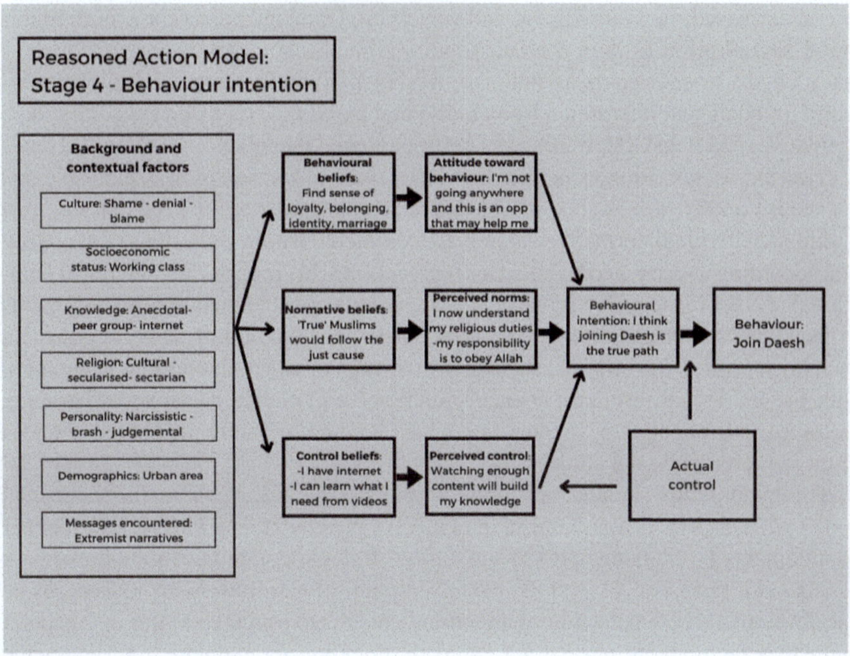

Fig. 4.8 Behavioural intention stage

on the subject matter to be able to make his definite choice, which, according to the model, is his behavioural intention to join Daesh, because he believes this is the 'true path'. Figure 4.8 below highlights how stage four of the RAM process takes X1 to the precipice of actualising his behavioural beliefs and behavioural attitude, alongside reconciling his normative beliefs and restrictions, with adequate behavioural control beliefs operating to keep him thinking he can achieve his end goal.

At stage four of the model, X1 has his mindset and aims aligned to the best of his 'ability' and control, based on his background contextual factors, exposure to the Daesh narrative, and a desire to do something he views as 'worthy'. Parents, friends, and others have no real tangible idea of his actions. At best, they may believe he is becoming more religiously aware, or at worst; they may be oblivious to the subtle attitude and behaviour shifts he has been demonstrating, as in many instances, such actors may not fully understand or show awareness of the role of social media and its possible impacts on behaviour. Up to this point in X1's journey, there has been an overload of pro-Daesh content, narrative and ideology consumed with no obvious or identified point for intervention or alternative to be added to the context. The process X1 has undergone has taken the foundation layer of socialisation, which can be identified as 'general socialisation', built a bespoke or 'specified socialisation' process (politicised socialisation, extremist ideas and hate-based narratives) on top of this foundation layer, and is now at the stage where the rise of Daesh and its communicative strength has created a 'situational socialisation' layer that consists of context-based 'new-normative' and value-based vantage points for action. In this guise, socialisation as both a process and a critical theoretical framework for this study begins to be seen as an iterative, incremental, and situational reality experienced, imbued, and interpreted by an individual based on a multi-layered attitudinal, normative, and behavioural milieu of self-negotiated meanings.

Viewing socialisation as neither linear nor fixed in transmission is plausible and possible. Equally, it is argued that multiple socialisations are in effect at any time within an individual normative value system, which positions iterations of socialisation according to want, need context and perception. This proposition has modal (situational needs), manageable (control needs), and measurable (action needs) elements to each layer of socialisation (general, specified, and situational). This suggests that each layer within this initial hierarchical structure is, in fact, in constant negotiation with the individual through personal preference and choice narratives. There are, therefore, self-negotiated, socially negotiated, and emotionally negotiated versions of these preferences and choices.

This hypothetical socialisation model requires a detailed analysis of the conceptual and applied uses of its assertions in the P/CVE CN context. In addition, how it either complements or challenges RAM within the strategic communications framework of this field. For now, X1 is at the perceived point of no return in his journey, as he has developed a core behavioural intention, which is to join Daesh. It is at this point in the hypothetical journey that AX enters the equation and domain of X1 with what is known as strategic communications-based CN. For AX's example 2 video (five considerations on Syria) to be compelling, it would need to affect X1's behavioural intention by causing him to reconsider elements of his perceived controls, norms, and normative beliefs, as well as impact his attitudinal and behavioural beliefs. Figure 4.9 below illustrates the introduction of AX into this process.

X1 is technically ready to join Daesh and only really needs to say this much in an online engagement with the right person to find himself introduced to another layer of online communications from which things become serious for him. Whilst searching for queries like 'how to join Jihad in Syria' on Google/YouTube, the AX example 2 video shows up as a match on X1's YouTube channel and, to a greater or lesser extent, takes X1 by surprise. He does not skip the video or end it and is taken in somewhat by the provocative thumbnail (static image associated with the video). He keeps watching and finds himself going through a raft of emotions and thoughts, which ends up with him needing to watch the video three more times back-to-back. The style and content of the video appeal to some of his general socialisation understanding of popular culture, specific socialisation through the use of language he recognises from other videos on the subject, and situational socialisation due to his decision to try and join Daesh. From a P/CVE strategic communications perspective, the content's message, messenger, and medium need to be in enhanced alignment regarding access, credibility, and some form of 'call to action'. AX's video would possibly only have one real chance to impact positively on X1 and could also easily have the opposite effect (galvanising existing beliefs). From a conceptual standpoint, the RAM approach has been seen as an effective process to seek to change behaviour in the field of health management and behaviour and has also been applied to P/

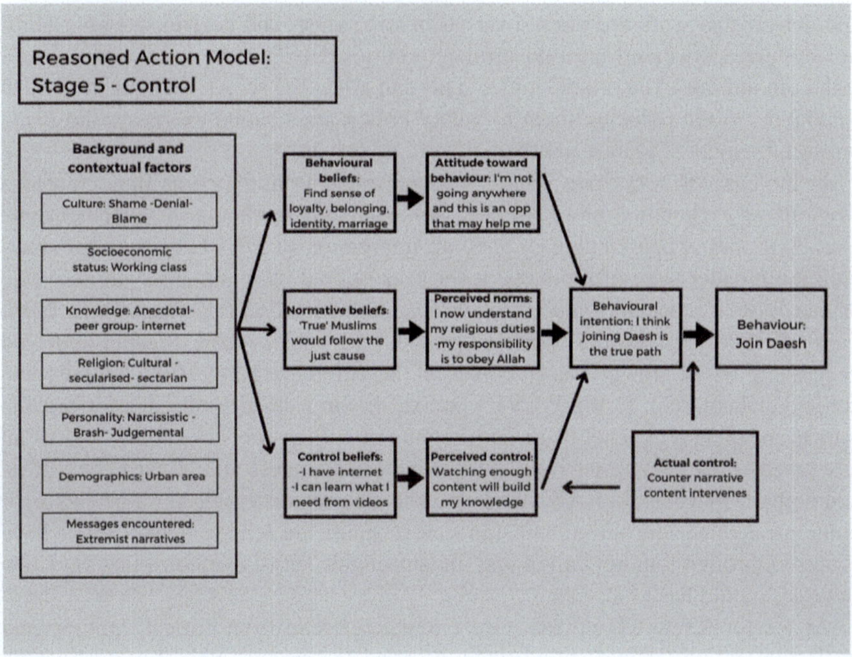

Fig. 4.9 RAM and counter narrative interventions

CVE CN efforts. However, its application as a model for behaviour change leaves out one fundamental and intrinsic element of the human condition and its effect on intention, attitude, and behaviour: emotion. This will be addressed in detail later, but for now, the discussion moves on to outline some of the real world impact of the AX campaign in terms of outcomes, success criteria, and initial learnings. This section should also help to facilitate a more detailed analysis of the P/CVE CN field later in this study.

The examples utilised in this case study are taken from the pilot campaign administered through RAN's 'Counter Narratives' working group, as it was known at the time. This group eventually morphed into the RAN Communications and Narratives working group, expanding its membership and scope to include 'best practice' and 'emerging tech' needs. During the pilot campaign, basic impact measurements were aligned with the vanity metrics approach to mapping 'success criteria'. This meant that terms like 'impressions' (how many devices the content was sent to), 'likes' (how many users pressed the like button), 'shares' (how many people shared the video via their channels or to other platforms), and 'reach' (the total number of people who see the content) were put forward as barometers of success for this campaign. The nature of the 'private–public' partnership model in this case was seen as a positive step in making the overall counter-extremism effort more joined up and credible. There is perhaps a note of caution in this example, too, in that despite coming across as a partnership model, the community sector organisations or grassroots entities that create and deliver this work are often dwarfed in size, scale, and decision-making abilities by huge networked intra-government entities (EU/DG Home) alongside giant tech corporations (YouTube/Google). This can affect the scope and context of such campaigns, often reducing them to vanity policy-justification exercises and using only vanity metrics to measure them.

In the case of AX, these issues were prevalent in as much as they created a dynamic of expectation along with an atmosphere of corporate management and marketing that, when combined, gave an impression of P/CVE being a 'product' within a broader suite of products or services offered by commercial entities who primarily seek monetary value through revenue generation (tech sector). In 2014, the difference between the 'organic' growth of a social media channel and paid advertising-based growth was perceived as the difference between authenticity and commercialism [15]. In the P/CVE context, this translated to the idea of organic growth meaning a channel or piece of content has grassroots credibility versus an ads-based effort being contrived and lacking the openness and transparency of its counterpart. What makes AX stand apart in this regard is not any one factor or outcome, in that some of the campaign's most significant achievements seem to have occurred through a more natural and uncontrollable set of dynamics that very few observers could have foreseen.

At the time, few CN efforts were circulating aside from critical thinking and resilience-building content that lacked dynamic and creative flair. Too many static websites closely aligned to government-funded efforts merely clouded the media environment and the conceptual approach to tackling extremist content online [11]. By way of contextualising the era of AX's rise, the US State Department's 'Think Again, Turn Away' campaign sought to use Daesh's content against them. The assumption here was that by appealing to the viewers' sense of moral and ethical scope, they (the audience) would 'understand' that terrorism is not the solution. The sheer openness of this intention is not the sole reason for this campaign's quick demise. It is relatively simple to offset the 'message, messenger, medium' model of effective CNs and show that the US government would lack the trust, credibility, and creative understanding to penetrate online users' mindsets with anything but irony. This merely becomes an extension of existing communication tools with which to aim at the government and its role in creating grievance and mistrust. Simply put, CNs during AX's time were not CNs at all. They had no theological, ideological, or positional legitimacy or credibility. The era in P/CVE was tantamount to telling diaspora audiences that 'terrorism is bad, so do not do terrorism'. This comes from the 'hypodermic syringe model' of classic media communications theory, which suggests communication is a one-way process from messenger to audience, who consume and accept the message.

The only known comparative to AX during this time would have been 'Burka Avenger'. This was a carefully administered animation-based campaign targeted at Pakistani audiences native to Pakistan and sought to address narratives around misogyny, domestic violence, and women's rights [17]. This campaign received support from the US State Department through grants and NGO links to Pakistan [18]. It is a well-known truism that the term 'NGO' in the context of Pakistan denotes an entity created with foreign support and resources, administered usually by Western-educated 'elites' of Pakistani society [12]. The one anomaly was the Edhi Foundation, a localised women's refuge and hospice service run by the now-deceased Abdul-Sattar Edhi [8]. The need to explain these points is twofold, firstly, it provides a contextual understanding of the operating environment into which AX entered. Secondly, it highlights an obvious flaw in P/CVE approaches during the sector's most formative years: lack of impact due to lack of understanding, compromised motivation, confusion, fear, or reluctance. There are too many variables to address this argument in this section, but later sections will allow for a more robust analysis of these issues and their implications for P/CVE. The emphasis in this section now shifts back solely to the outcomes of the AX campaign that went beyond the commercial understanding of success in communications and from which a conceptual and practical set of learnings emerged regarding CN communications. It is prudent to include performance metrics to contextualise the results more holistically. To this effect, the format will break down dissemination tactics, engagement tactics, outcome criteria (known and unknown) and performance measurement. This structure will enable specific elements of the campaign to be highlighted to apply the theoretical framework where relevant.

Strategic Communications and Dissemination Considerations

The official pilot phase ran for over six weeks between March and April 2014. In this period, targeted 'ad campaigns' were set up and run across YouTube and Facebook platforms. A vital component of the approach was to apply the A/B testing model, with the same content delivered to different target audiences through fairly basic audience segmentation criteria demographics-wise and more targeted criteria as a variable sample. Despite the audience targeting criteria being relatively basic and focusing on broad points of segmentation (age, gender, location) to look at user 'interests', this criterion enabled A/B testing to be applied in a traditional method. It facilitated previously unused methods of audience segmentation to test the content, platform performance and flexibility, sentiment tracking and generation, and furthering the knowledge base of P/CVE-centred targeting needs in the online space. This review of the performance elements of the campaign begins with a look at the dissemination tactics employed.

The choice of YouTube and Facebook had, at the time, specific rationales behind them, which would now lead to other platforms such as Instagram, TikTok, Discord and more being included. In 2014, two apparent needs concerning the choice of these platforms arose. Firstly, the vast majority of pro-Daesh content was uploaded and then shared on YouTube, which had, by this time, become the go-to source of user-generated video content globally. Secondly, Facebook was also a key platform for narrative spreading, instant messaging and content sharing during this time.

Private Facebook groups were commonplace [9] and meant that extremists could share content and ideas in relative privacy, away from public-facing communications. They could also invite those they felt may share their ideological leanings to these groups. The nature and style of the AX content (animated videos) meant that YouTube was the ideal platform to disseminate content, with Facebook serving as a way to serve ads to attract users to the AX content on YouTube. This decision progressed the dissemination needs onto the issue of who to target and to use which criteria.

During this time, the social media marketing space depended heavily on linking targeted ads to Google search engine results, the Google display network, and YouTube ads services. The search engine results were based on providing ad-based prompts to users by tracking their search terms entered into the Google search engine database, from which algorithms would display ads that included similar keywords and hashtags. The display network version used algorithms to place relevant ads similarly, the key difference being the display of ads on third-party websites that may have had broad links to search terms that matched the ad targeting criteria. The YouTube version used algorithms to match user searches on the platform to advertisements that were again similar, using hashtags and key terms. These ads would then appear on a user's YouTube account or channel as prompts that could be selected. They could also appear as short ads that played before the specific content that a user had chosen to watch. Alongside this technical architecture, a more campaign-specific marketing approach was also needed to drive content to intended audiences.

The specific marketing approach for AX had three layers designed ultimately to end with some form of positive engagement with the content. From the reasoned action (RAM) approach, this relates to an intervention that occurs at the behavioural intention and behavioural control stage of the model. This is where it counters the narrative and intended behaviour of the individual. The subject is either becoming more comfortable with watching extremist content or has decided to support or join Daesh through being taught the ideology and cause being offered by the network. To achieve this form of online intervention, the three layers mentioned above corresponded to creating awareness of AX among the target group, creating an element of influence on 'traffic' that expressed positive sentiment towards 'Jihad' or a likelihood to reach such content, and then to engage the target group with AX content. This process and layering of stages in the communication versus influence context is illustrated below in visual form (Fig. 4.10).

The first stage of this funnel is exposure, which entails target audiences seeing the AX brand advertising. As the targeting criteria have been based on utilising an essential set of terms, searches and online platforms (websites) that match types of search and content associated with relevant issues, this first level of the funnel gets the AX ads in front of the necessary audience to a greater degree. Stage two of the funnel is the base-level influence that watching the videos by clicking through the ads creates. At this stage, it cannot be known if this initial influence will result in a positive action campaign objective-wise. The third stage of the process is engagement, which relates to audiences who have seen the content and then made an active choice to engage further through vanity metrics such as likes and shares and subscribing to

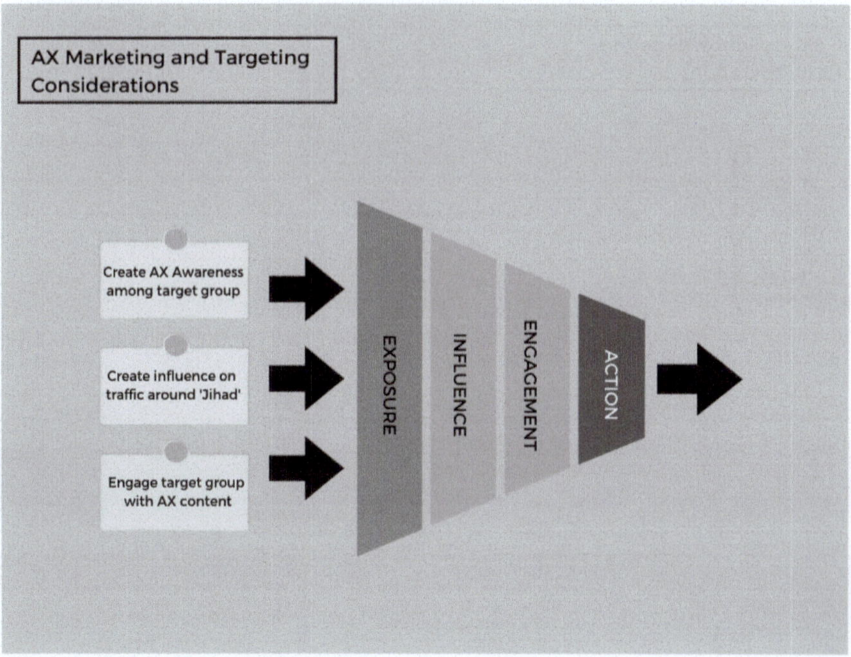

Fig. 4.10 AX marketing funnel approach

the AX YouTube channel. The final stage would be the action (rejecting extremist narratives or engaging in more positive pathways). Reaching a stage where audiences express sentiments (both positive and negative) is naturally an outcome that any P/CVE CN would seek to achieve. Such reactions suggest the content has elements of effective strategic, narrative, and delivery features that, when consumed, warrant action through sentiment creation. This leads to another aspect of the campaign process, which relates to how effectively (or not) a CN product both targets and understands the types of sentiment its intended audiences may subscribe to personally, publicly, and ideologically. The appropriate strategic communications name for this requirement is audience targeting and segmentation.

Understanding audience needs at a granular level is essential for a campaign created specifically to undermine and challenge the basis of the narrative of Daesh and Jihadi ideology more broadly. A strategic intention to prevent individuals from seeking to support or join Daesh requires such depth. For the AX campaign, the potential audience spans those curious about the situation in Syria, those curious about Islam, those interested in learning about Jihad, and those actively considering leaving the UK for Syria to join Daesh. These varying degrees of need and intent can be distilled into four categories or phases of attitude and behavioural belief development illustrated in Fig. 4.11 below and best understood as different types of mindsets associated with the issue.

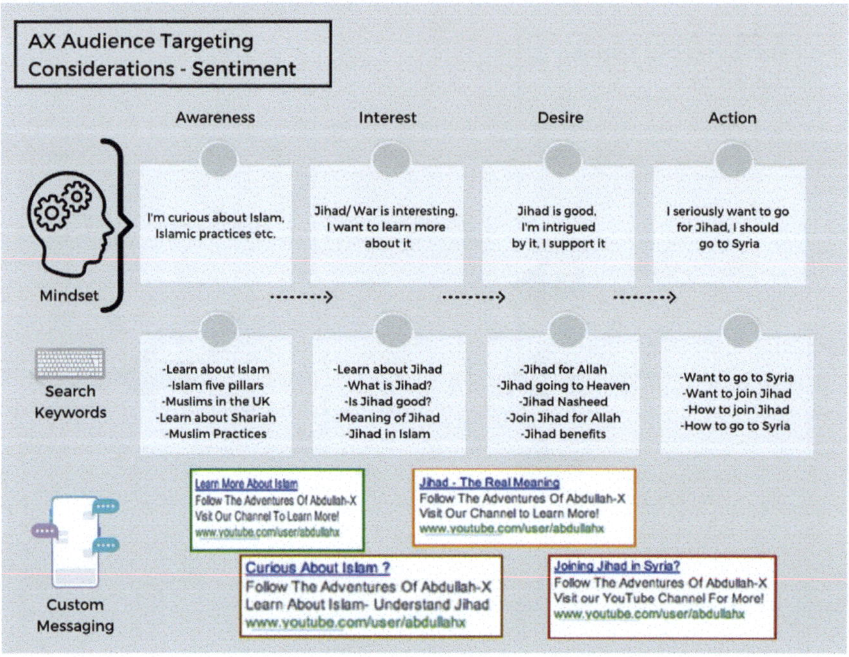

Fig. 4.11 Targeting, sentiment and custom messaging

The illustration above points to an intrinsic link between creating effective ads and online searching for various mindsets, some of which are more vulnerable than others. It is feasible to suggest that an online user who is displaying nothing more than curiosity and feels the internet is the appropriate place to feed this curiosity could find themselves slowly, or quickly even, being taken down the path of engaging and seeking more and more serious content. The rationale behind this point is that during such a process, there is little evidence to suggest any behavioural controls are in place to limit engagement with extremist content. This means the developing behavioural beliefs and attitudes are not subject to 'checks and balances', as the online environment lessens the extent to which normative beliefs and controls would come into significant consideration. To raise awareness of a product (AX) designed to intervene at the behavioural control stage, the type of search keywords used need to be used to offer up custom AX ads that can appear pretty high up on the search results displayed to the user. In the AX context, Fig. 4.12 below highlights how this would look in a search results field after a user enters 'learn about Islam' into the Google search engine.

The illustration above brings up a custom AX ad on the right-hand side (green highlight) and offers the type of short narrative that may appeal to specific individuals within the target group. This approach may not be practical for those already well down the path of being brainwashed by Daesh ideology. Still, those who are curious or seeking attention may find the narrative of the ad exciting. This approach can,

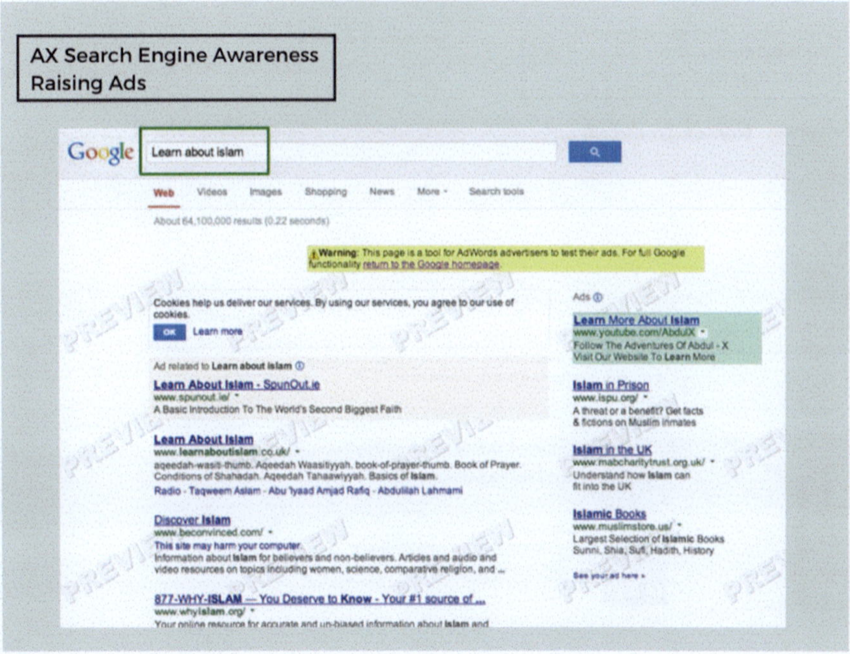

Fig. 4.12 Search engine results—raising awareness of the AX brand

however, intervene at another stage in the process, which in this case would be the 'action' stage. Suppose a user is actively searching for information on a potentially harmful action. In that case, another more robust and highly provocative custom ad may divert this search to the path of AX content. As such, should the user engage with the custom ad they have seen on the display results, there is a chance this individual may end up watching AX content, creating the scope for presenting an alternative view. From the innocuous 'learn about Islam' search question to an altogether more serious line of enquiry, raising awareness of the AX brand still offers a constructive route to seek engagement opportunities with individuals caught up in behaviours that can lead to harm and danger. Figure 4.13 below illustrates this point accordingly.

As mentioned earlier, through search results, display network ads, and YouTube ads, the AX campaign can build awareness among different users, each with different needs and motivations. Display ads on third-party websites are known in the social media ads industry as keyword contextual targeting. This provides an ad campaign with space on websites that includes terms and contextual data that match the ad campaign's database of key terms. For the AX campaign, it is evident that a comprehensive set of key terms and specialist criteria were proposed, given how much engagement the campaign garnered over time. This is something that will be explored shortly alongside identifying the target audience demographic, but at present, it is helpful to demonstrate how this element was manifested in the actual campaign. Figure 4.14 below highlights how contextual keywords were applied.

Fig. 4.13 Action stage—raising awareness of the AX brand

In the above example, keywords used in the ad campaign process were 'Islam', 'Dawah' and 'Allah'. The AXE ad banner has been displayed as these are also present on the webpage 'Islamawareness.net'. The use of the term Jihad in the ad can prompt and engage users to click on the display ad, with the link then presenting an AX video. The third method of targeting used the YouTube ads service to bring up AX ads on videos users watch. During the AX pilot campaign, the brand attracted much media attention, which the campaign sought to use to its advantage, as per below (Fig. 4.15).

The above example shows an AX YouTube ad at the start of the Vice News documentary on British Nationals fighting with Al-Qaeda in Syria. The AX ad plays, and the AX channel is also displayed on the right hand of the screenshot. This type of ad display on content could effectively drive someone to the AX channel, even if it is out of curiosity, having seen the Vice News video. This builds on the dissemination, engagement, and ads tactics, offering helpful insight into how the campaign was set up and run. To complete this review, however, it is essential to outline the intended audience. Knowing this will help to gauge how well the campaign performed and if any unexpected results also emerged. In Fig. 4.11, a visual representation of the targeting criteria was presented. This links to who the messaging within AX content was intended for, known as the recipients. AX audience targeting was seeking two broad categories in this regard. The first target audience was individuals developing increasing levels of curiosity regarding the rise of Daesh in the global information, geo-political and everyday discourse environments. This curiosity level could range

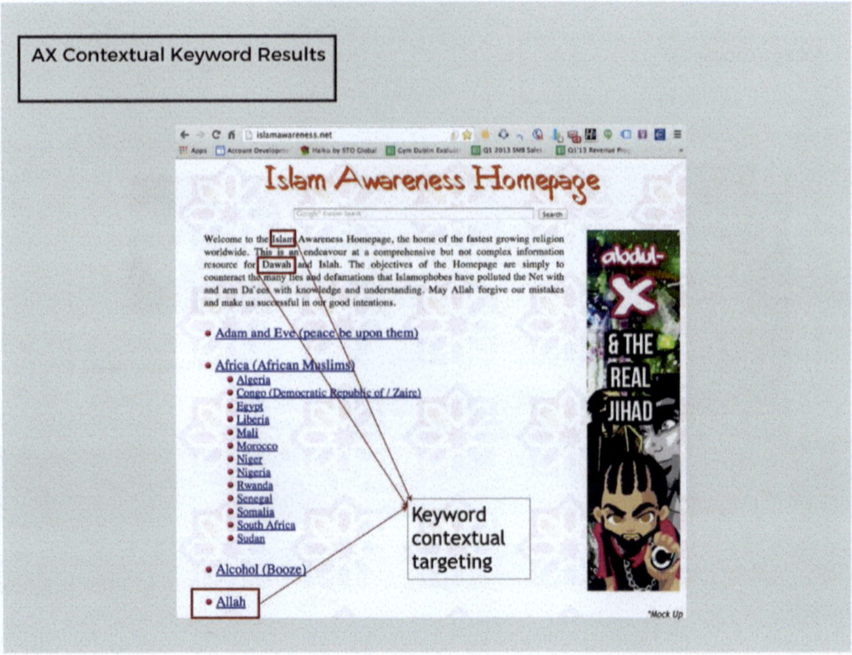

Fig. 4.14 Contextual keyword term display ads

from having contact with offline discussions, online engagement with news media and social media content and even initial contact with extremist narratives (knowingly and unknowingly). From the RAM model perspective, this target group ranges from those with developing behavioural beliefs (seeking out belonging, identity, acceptance, etc.) to those inching towards possessing a specific behavioural intention (seeking to join, promote and support).

In this context, the AX narrative would seek to affect normative beliefs and the control beliefs associated with these norms to intervene in the perceived normalising of extremist thinking and the amount of actual control an individual possesses over-taking such drastic action. Within normative beliefs, socialisation processes become active concerning the specific background and contextual factors relevant to each individual's journey up to this point (coming into contact with AX content). Therefore, socialisation's specific 'agents' are critical facilitators of the propensity, intensity, and inculcation level of positive or negative normative beliefs. The link between agents of socialisation, value alignment, human agency, and action come together in strategic communications and behaviour change contexts under the broad auspices of influence and persuasion. These terms require detailed discussion around their applicability and dynamics in the digital era, the historical timeframe explored in this case study.

The second target audience is those with a firm sense of behavioural intention who actively seek ways and means to action this intention. Such intentions could range

Fig. 4.15 AX YouTube Ad serving

from making travel plans, thinking about what acts would suffice as a show of support, or that they want to champion the cause through promoting it. AX content seeks to intervene at the behavioural control stage, offering the individual a short, sharp, persuasive shock to their cognitive information-processing system or mindset. The RAM element, thus, is looking to reverse both normative and behavioural alignment and offset the impact of previous socialisation(s) with a newly rendered socialisation. This would be delivered within the CN and can be labelled' re-socialisation' at this stage of this research effort. This is a somewhat basic premise that it is hoped can be further developed in the conceptual framework proposed towards the end of this book once the analysis is complete. The use of the English language, cultural references, and specific iconography also point to the audiences being UK-based and aged 14–35 years. The gender breakdown suggests a leaning towards men being the primary group. The content is narrated by a male voice, which supports this emphasis.

The examples of audience targeting, marketing, and promotions presented here point to a process-driven, commercial formula for how products, brands, and services can reach intended target audiences. As such, within a P/CVE context, such practices demonstrate an approach that seeks to harness tech tools in the digital space to reduce the likelihood of individuals choosing to act upon extremist ideas. Critical conceptual discussions emerge from this context, not least from the perspective of reasoned actioned theory and socialisation. In essence, these two theoretical frameworks possess sufficient depth to be able to unpick design, delivery, and outcome elements from within P/CVE CN efforts, simply due to being approaches highly related to behavioural change examination (RAM) and normative values generation (socialisation).

Using the case studies in this research project to present an overall analysis of how P/CVE CNs seek to change behaviour and reduce/prevent violent extremism allows nuance and process to be elucidated in a 'lessons learnt' context. This can include the proposition of both strengths and weaknesses or limitations of the CN approach and examining the reliability of measurement outcomes of such efforts, which links to how replicable campaigns and their design processes are if used again in a different context. The issue of critical outcomes for the AX campaign is something that can now be presented to build the descriptive and analytical foundations for a robust analysis of P/CVE CN impact later in this work.

As a short-term trial campaign, the project assumed the characteristics and dimensions of a high-intensity 'quick-win' effort, commonly observed in commercial mainstream campaign situations, to have an immediate impact but not necessarily longevity. Despite being a product initially produced at the grassroots level with no outside influence, financing, or in-kind resourcing, AX rapidly transformed into a fully functioning and tech-centric 'collaborative' endeavour that harnessed the scale and velocity of a global mass marketing campaign. Using advanced (for the period in question, 2014) algorithmic tools to target, pinpoint, and deliver messages to individuals at various points along the path of extremism, violence, or ideological indoctrination, there is little doubt that the campaign delivered cutting-edge communications on multiple levels. It is evident from this case study discussion that AX integrated highly nuanced, culturally and ideologically/theologically appropriate narrative arcs, packaged them into unique animated videos and distributed them with the effectiveness of high-value corporate-level strategic communications activities. To evaluate the overall effect of this effort, it is necessary to have a thorough understanding of the campaign's primary outcomes, which give context and permit analysis of intended and unforeseen impacts.

Campaign Objectives and Outcomes

In terms of the overall objectives of the campaign, taking into account the formulation of a strategic framework for the communications design, and as previously highlighted in detail earlier (Fig. 4.3) in this case study, AX sought to produce the following strategic objectives:

1. Increase audience resilience levels
2. Build awareness levels of audiences
3. Expose flaws in the narrative and links to ground reality
4. Degrade adversary value propositions

These overall objectives can be broken down into two audience-centric objectives (1 and 2) and two narrative/adversary-centric objectives (3 and 4). Therefore, it can be inferred that AX delivered both prevention/inoculation-style intervention and refutation/undermining-style intervention synergistically as part of a customised CN package with the capacity to reach diverse audience profiles and needs. As is customary in online strategic communications practice, success metrics are derived from data such as views, reach, viewing retention, engagement, likes, and shares. These 'vanity measures' provide ready-made quantitative data that infer relative 'successes' of online communications from a broad perspective, implying that a campaign has had some impact in the digital realm. The issue with using this approach in the context of P/CVE is that such measurements and what they imply should ideally be linked to behaviour change dynamics, which may be captured to differentiate between efficiency and efficacy. The former is the domain of partners and stakeholders, budgets and public relations. In contrast, the latter is dependent solely on data that can demonstrate a positive change in the behaviour of an identified audience concerning a reduction in, rejection of, or inoculation against extremism/violent extremism. This discussion and argument are further complicated by the reality that the P/CVE industry (online and offline) is still in a learning phase and cannot claim to be at the pinnacle of its art, understanding, or delivery. This predicament is exacerbated by the constant conceptual, empirical, and moral debates that surround P/CVE, including the usefulness of radicalisation discourse, the use of psychology over ideology in interventions, the ambiguity of intervention objectives, and a general attitude that seems to isolate P/CVE efforts and relegate such issues to policy and academic circles rather than integrating them into a more prosperous, more socially relevant environment. The extent and severity of this topic will be addressed later in this research. Still, it is crucial to highlight that the outcomes of the AX campaign that will be examined next necessitate a nuanced analysis when this subject is contextualised in this manner. Regarding the overall results of the six-week pilot campaign, AX delivered the following outcomes (below). These are based on the 'Five Considerations for a Muslim in Syria' video and 'The Real Meaning of Jihad' video, alongside some promotional videos and video shorts. At the time, these results were considered to be a measure of success for a P/CVE campaign (Fig. 4.16).

To make more detailed and contextually relevant use of these metrics, the measures are defined as:

- Impressions: Total number of devices the messages were delivered to
- Video Views: Users who consumed thirty seconds or more of a specific video
- Channel Subscriptions: Number of users who subscribed to the AX YouTube channel
- Comments: User engagement through leaving positive/negative comments on videos

In addition to the vanity metrics outlined above, some of the tools at the disposal of YouTube partners allowed for the tracking of user engagement across the campaign in terms of content viewed during the same timeframe as AX content was being pushed to target audiences using the various search criteria described earlier in this chapter. Even though this method yielded some intriguing results, they were based solely on two distinct user journeys (only user journey 'A' is most relevant to the

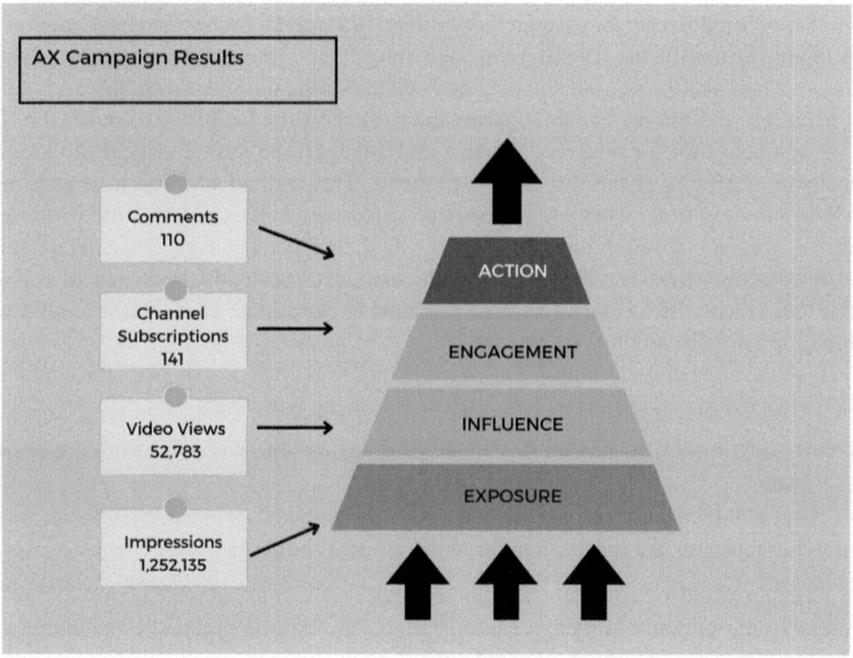

Fig. 4.16 AX campaign results

aims of this study). They thus cannot be considered representative of a broader trajectory regarding behaviour change outcomes. It is prudent to suggest that, while such data can certainly add credibility (validity) to a campaign's results, ensuring that such results can be replicated remains a significant challenge for online P/CVE initiatives. In the conclusion chapter of this research project, some of the primary causes of these problems will be examined.

From the perspective of the theory framework, however, these two encounters with the AX campaign allow for a greater understanding of how the reasoned action approach is applied to strategic communications efforts of this nature and how its conceptual assumptions manifest in the operational environment. These two examples also enable more process and normalisation dynamics to be added to the role of socialisation and agents of socialisation concerning online and social contexts in the offline world. In particular, such user engagements and subsequent online journeys of exploration present an interesting context through which different types of socialisation processes can be conceptualised and begun to be better understood, rather than the current P/CVE reliance on the highly contested concept of radicalisation to be continually rolled out as the 'answer' to the extremism problem.

User journey 'A' links to offline engagement to supplement the online efforts through workshop-based sessions designed to raise awareness of the AX CN offering and build interest in exploring issues linked to critical thinking, decision-making, and online safety. Within the context of learning more about online safety, a small number of 'digital resilience' sessions were held in secondary schools in East London, engaging primarily the 14–16 years age range. At one such session, a student engaged positively with online safety and source verification ideas when searching, consuming, and sharing content. During the workshop, the facilitator mentioned AX as a valuable tool for students to learn and engage more about critical thinking, decision-making, and the threat of extremism. This individual went in search of AX in his own time. These 'independent' research results yielded some exciting results from socialisation and reasoned action theory perspectives, respectively (to be discussed shortly). Having found the AX channel, the individual watched all available AX videos, 'liked' them, and felt the need to comment "great videos" and say that others should watch them too.

A short time before finding AX, this individual was engaging in YouTube content that when examined closely, offered up the following narratives:

- "These are the Mujahideen that they are accusing. Jihad is one of the doors to heaven."
- "God will take your sadness away if you follow Jihad."
- "The state of Islam has been built on the blood of honest Muslims."

The "great video" comment was about a teaser AX video in which the AX character tells the audience fairly aggressively about various issues related to the internet, identity formation, and faith, as shown below in the extract.

"*You will need to ask yourself, who is it that you need acceptance from? If you think this has something to do with Islam, this internet identity, then at least seek out knowledge before assuming that being called a Jihadi is somehow based on the traditions of this faith (Islam). Identity is about character, not false impressions." (Abdullah-X, Teaser Video, 2014).*

In the space of a few short weeks, this individual seemingly moved away from consuming overtly extremist and pro-Jihadi content to watching exam tips and study skills videos, an example quote of which is below:

> *"Considering it's exam season, I wanted to tell you all about how you can smash your exams."*

Analysis of AX Using the SoRaSCo Integrative Model

At a purely conceptual level and as the basis of this research project, this user experience facilitates avenues of exploration regarding RAM, strategic communications approaches, and socialisation theory. From this vantage point, this individual's engagement with AX-related content and narratives points to an 'intervention' by the AX brand that could be classified as addressing a 'known unknown'. This relates to the issue that it was a 'known fact' and accepted position that the Daesh narrative was targeting young people and were also being made value propositions about the relative merits, rewards, and associated benefits of undertaking Jihad, leaving their homelands and accepting the 'cause'. What was not known, or less evident, would have been who these young people were, where they were and how they were being engaged with by pro-extremist content at an individual level. As such, they are, in essence, 'known unknowns'.

In this form of online strategic communications intervention, anomalies remain regarding pinpointing specific background and contextual factors, which does cause some reliability issues for the RAM approach. However, at the stage of behavioural belief formation, normative belief development, and control beliefs of the RAM model, an intervention presented in this example can already be seen to be starting to redirect and repurpose the individual's beliefs, norms, and controls. The offline encounter with AX content leads to a process of engagement that results in the attitude towards behaviour, perceived norms, and PBC becoming realigned, thus changing the behavioural intention and, therefore, the behaviour itself. This individual goes from engaging with pro-Jihadi content to engaging with exam technique videos. This process is precisely why a model operationalising socialisation theory appears to be essential to understanding the uses of CN communications. As such, the following section delves into the links and uses of this approach.

As noted earlier in this case study, the RAM approach links on many levels to socialisation theory. In particular, the processes by which an individual takes on behavioural beliefs, attitudes, and perceived norms align heavily with the fundamentals of socialisation. With a CN product such as AX, the emphasis is on offering challenges but in a supportive and considered fashion. CNs seek to replace harmful ideas with mainstream or non-violent pathways in challenging problematic norms and attitudes. This process is, without doubt, a process of socialisation, or re-socialisation, to be precise. The RAM approach charts a systematic process from an individual's background and social context, then connects these with distinct behavioural and motivational cues in the build-up to possible action. Socialisation theory can play an essential role in adapting the RAM process by including clear parameters that identify which norms/values are problematic and which norms/values can replace them. The task of the communications angle in this context is to create robust, credible, and engaging content that is able to provide the need, motivation, and access to re-socialisation paths upon acceptance of the narrative's overall message. To understand the overall role and links between behaviour change models (RAM), social dynamics and norms (socialisation) and communication (strategic), as applied to this case study, the following table charts each element (Fig. 4.17).

The table above suggests that socialisation facets reside at the core of the RAM approach to behaviour change and strategic communications considerations. The theory offers insight into how specific background factors can affect individuals in different ways, the role of socialising agents in developing new perceived norms and attitude formation, and how a reversal of one such process requires a form of re-socialisation. As mentioned previously in this study, socialisation theory has limitations if used in isolation, especially concerning needing help to chart individual versions of how norms and values can be manifested. However, political socialisation theory does go some way to distil such manifestation within a lens of change, transformation, group loyalty, and ideological alignment. Equally relevant to this case study, the combined use of behaviour change models and effective communication

Analysis Using 'Communicative Socialization' Model - AX Case Study

CRITERIA	AX PROJECT TACTIC	SOCIALIZATION CONTEXT	COMMUNICATIONS CONTEXT
BEHAVIOURAL INTENTION	USE OF EMOTIONAL AND LOGIC BASED DRIVERS TO ELICIT SELF-REFLECTION	MAKE USER RECONSIDER SOURCES OF INFORMATION AND INFLUENCE	NARRATIVE DRIVEN CONTENT ADDRESSING INDIVIDUAL NEEDS
ATTITUDE TOWARDS BEHAVIOUR	USE OF RISK VS HARM NARRATIVES TO CAUSE DOUBT	REINTRODUCE NORMS AND VALUES AROUND CHOICE AND DECISION MAKING PROCESSES	NARRATIVE DRIVES BUILDING OF KNOWLEDGE AND AWARENESS
PERCEIVED/ SUBJECTIVE NORMS	CREATE DISPARITY BETWEEN PERCEIVED NORMS AND ACTION	CHALLENGE CURRENTLY HELD NORMS/VALUES AND THEIR ORIGINS/VALIDITY	NARRATIVE ENCOURAGES A RETURN TO MAINSTREAM VALUES
PERCEIVED BEHAVIOURAL/ ACTUAL CONTROL	ENCOURAGE CRITICAL THINKING OF PERSONAL 'OBJECTIVES'	OFFER ALTERNATIVE PATHWAYS TO VIOLENCE THROUGH NEW MINDSET	NARRATIVE OFFERS USER A CHANCE TO CONTROL ACTIONS THROUGH COGNITIVE CHANGE
BACKGROUND CONTEXTUAL FACTORS	CULTURAL, RELIGIOUS AND SOCIAL DYNAMICS RELEVANCE	OFFER ALTERNATIVES TO HATE AND EXTREMISM VIA SOCIAL ACTION	COGNISANCE OF EVERYDAY SOCIAL REALITY WITHIN NARRATIVE CONTENT
FEEDBACK & ITERATIVE APPROACH	USE OF USER FEEDBACK TO INFORM NEW CONTENT	MAKE NARRATIVE RELATABLE TO SOCIAL REALITY & CREDIBLE	USE FEEDBACK TO IMPROVE CONTENT DESIGN

Fig. 4.17 AX and SoRaSCo model analysis

tactics appear to have given socialisation-based needs an ideal foundation to build. With the added element of a clear intent for feedback and an iterative approach to such communications, this model of analysis has been applied effectively in a case study context. Some outlier issues and broader points regarding the AX project warrant discussion. Some of the project's more unique outcomes are worth discussing and can be found below.

Broader Topics of Interest to the AX Case Study

A particular 'anomaly' identified during the AX campaign was that the CN presented by the AX character found its way to the consumption needs of a user in no identified target audience group. The case in question (user B) relates to a middle-aged white male who came across AX content and engaged with the channel. This individual had previously engaged with anti-Islamic content and 'liked' such narratives. During this period, the user also encountered the AX video 'The real meaning of Jihad', leading the user to feel the need to comment on the narrative and its message, saying the video was "thought-provoking". The same user then revisited the AX channel and watched another short video addressing those who follow the path of hate and call this 'faith'. The user leaves another comment, "Very good, makes sense". These two examples of different users from different social backgrounds, one being part of the intended target audience and the other being someone who is unintentionally targeted by the content through virtue of engaging in other content that links to AX content, serve as specific interventions that offer some exciting angles of analysis from a theory framework perspective beyond the RAM approach, taking the discussion into the realm of socialisation theory and strategic communications considerations.

Online strategic communications efforts such as the AX project appear to be relying on a reasoned action approach that uses the facets of socialisation theory and communicative action to create the conditions for an individual to engage, consider, question and then reflect on narratives through a process of sharing implied meanings, vague cultural references and symbolism, and direct communication to elicit a response. This response is designed to be triggered through the activation of a new layer of PBC on the part of the individual that alters cognitive and behavioural dynamics in favour of the intended strategic communication (CN), thus resulting in a reasoned decision to perform a new action (reject extremist thinking etc.). This becomes more salient as an argument within this context of P/CVE when the next feature of the AX campaign is outlined.

The AX campaign sought to attract as many online users as possible and, therefore, resembled a mass communications model, albeit with a particular remit. The design of the narrative elements of AX content was solely based on a need to engage people who may be starting to become interested in extremist narratives. As such, AX was trying to intervene before extremist ideas were adopted. As mentioned previously, it can be argued that AX was also. As a result, they were trying to inoculate audiences to recognise extremist content if they encounter it and hopefully be able to reject it. There is nothing to suggest in the design elements, narrative content or alike, that AX communications were robust or challenging enough to directly address those who had already decided to take the path of extremism and were willing to use violence. How well would a few short, animated videos compare to a sustained, personalised, and human-facing set of inputs from Daesh affiliates? Assuming a two-minute AX video could match months and gigabytes worth of pro-Daesh content, private messaging and support is conceptually improbable. This makes the subsequent outcome of the AX campaign even more bizarre. This matter is being discussed in this guise because two other anomalous outcomes manifested during a fascinating turn of events for the AX campaign. These cast an even more curious set of considerations for the campaign, P/CVE learning, and the role of strategic communications in both.

The first was a detailed 'refutation' of AX's 'Five Considerations' video from someone representing Daesh in Syria. The 5,000-word (approx.) document spelt out why the points made in the AX video were 'wrong' and 'un-Islamic' and should be treated as such. Shared on the platform 'justpasteit.com', it appears the refutation was intended for active Daesh supporters to ensure they do not feel confused by the points the AX video makes. Below are two Twitter posts referring to the AX video and its refutation, titled '5 questions' (Figs. 4.18 and 4.19).

Being able to compel active Daesh supports in Syria/Iraq to engage with the AX content and share what appeared to be rushed and nervous refutations of a cartoon character's two-minute video was not something overly apparent in either the content

Fig. 4.18 AX refutation 1

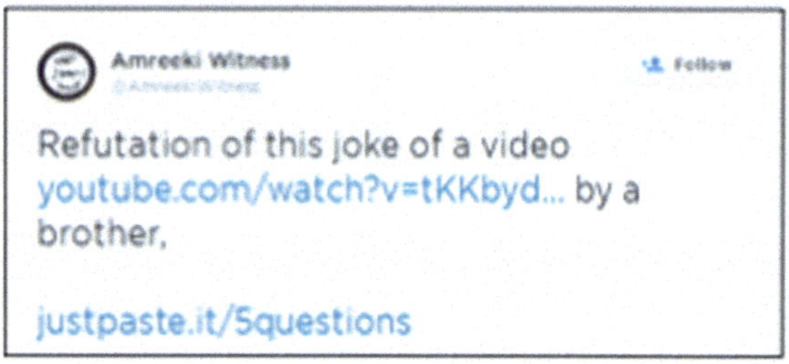

Fig. 4.19 AX refutation 2

design or strategy for AX as mentioned previously. What small windows of analysis that can be ascertained from this outcome are that communications created with a clear and specific goal, that have a narrative arc built from solid ideological and cultural positioning, with the power of mass marketing techniques and tech tools behind them, may yield far greater sets of outcomes than first planned for. The AX video in question comes across as a narrative delivered from a Muslim perspective, directed at Muslims and youth in particular. The combination of relatability, real-world critical thinking elements and packaging in a popular culture format have allowed the video to travel in the online sphere far beyond where it was intended to be consumed (UK geography).

What is even more intriguing is that the Twitter user in Fig. 4.19 'Abdallaah' aka 'Mujaaahid4life', eventually made contact with the AX channel and disclosed that he had left Syria and no longer supported or subscribed to the Daesh ideology. He also requested to meet with the video's creator, which was assessed as not feasible for obvious security reasons. 'Abdallah' turned out to be a Finnish national who had left his homeland to join Daesh in Syria and was significantly affected by the logical flow of the AX video. It is relatively safe to assume that the ground-level reality he encountered in Syria matched more of what AX described and not the glamorised dream world for Jihadis that Daesh presented in its communications. In this example, although very uncommon, the intervention occurs after the negative/harmful behaviour has already happened. This suggests that the RAM model can be adapted and expanded to consider 'controls' in strategic communications effective in broader contexts of withdrawal, realisation, and reflection processes after behaviour has occurred. It also points to the potency of socialisation theory in explaining how an individual can re-align himself with norms and values perhaps previously held and understood. The example suggests the process of cognitive dissonance when social reality does not match expectations and perceptions. Combined with rational thinking, it can affect a decision-making process for the positive in even the most 'extreme' and non-mainstream context (warzone).

It is impossible to delve deeper into just how much 'effect' or impact AX had on such an individual or if this could be replicated. Still, the example is one of the most unique outcomes of any P/CVE CN effort that has made it to the public domain. This is also the case for the other examples presented in this chapter, as they were unique journeys, and the campaign did not include a qualitative evaluation that would have found more users and sentiments to build a large sample set of data for analysis as to how the content impacted or affected users within the target audience. However, it allows this case study to conclude with an assertion that AX proved to be valid and effective in some specific contexts, aiming to provide credible CNs to audiences also being targeted by Daesh. The conclusion chapter will seek to broaden this debate with relevant contemporary threads, tying together theory frameworks and lessons learnt from subsequent CN efforts since the AX campaign.

References

1. Abdullah-X. (2015). *Abdullah-X - YouTube.* [online] www.youtube.com. Available at: http://youtube.com/abdullahx [Accessed 19 Mar. 2024].
2. Aftab Ahmad Malik. (2001). *The Broken Chain.* Amal Press.
3. Bauman, Z. (2006). *Liquid fear*. Polity.
4. Beck, U. (2009). *World at risk.* Polity.
5. Braddock, K. (2020). *Weaponized words: The strategic role of persuasion in violent radicalization and counter-radicalization.* Cambridge University Press, Padstow.
6. Brim, O. G., & Wheeler, S. (1976). *Socialization after childhood*. John Wiley and Sons.
7. Danziger, K. (1971). *Socialization.* Penguin Books.
8. Ehdi. (n.d.). *Management Style—Edhi Welfare Organization.* [online] Ehdi. Available at: https://edhi.org/management-style/ [Accessed 3 Mar. 2022].
9. Frennett, R., & Dow, M. (2015). *One to One Online Interventions—A Pilot CVE Methodology.* [online] ISD. Available at: https://www.isdglobal.org/isd-publications/one-to-one-online-interventions-a-pilot-cve-methodology/ [Accessed 20 Mar. 2022].
10. Giddens, A. (1990). *The consequences of modernity.* Polity Press.
11. Glazzard, A. (2017). *Losing the Plot: Narrative, Counter-Narrative and Violent Extremism.* [online] International Centre for Counter-Terrorism - ICCT. Available at: https://www.icct.nl/publication/losing-plot-narrative-counter-narrative-and-violent-extremism.
12. Hashim, A., & Reuters (2015). *Pakistan Announces New Policy for Foreign Aid.* [online] Reuters. Available at: https://www.reuters.com/article/idUSKCN0RV5D5/ [Accessed 7 Jan. 2022].
13. Howard, M., & Paret, P. (1984). *Carl Von Clausewitz On war*. Princeton University Press.
14. Institute for Strategic Dialogue. (2016). *The Impact of Counter-Narratives.* [online] ISD. Available at: https://www.isdglobal.org/isd-publications/the-impact-of-counter-narratives/#:~:text=ISD [Accessed 19 Nov. 2022].
15. Keen, A. (2012). *Digital vertigo.* Martin's Press.
16. McIntyre, L. (2018). *Post-truth.* MIT Press.
17. Mohsin, S. (2013). *Tackling religious intolerance and violence in Pakistan.* [online] CNN. Available at: https://edition.cnn.com/2013/09/24/world/asia/pakistan-christians-mohsin/index.html [Accessed 6 Dec. 2021].

18. Strong Cities Network and ISD. (2017). *Global Summit Denmark*. [online] ISD. Available at: https://www.isdglobal.org/wp-content/uploads/2017/06/SCN-Brochure_FINAL.pdf [Accessed 3 Oct. 2022].
19. Tzu, S., & Nylan, M. (2020). *The art of war*. W. W. Norton & Company.
20. Venhaus, J. (2010). *Why Youth Join al-Qaeda*. [online] Available at: https://www.usip.org/sites/default/files/resources/SR236Venhaus.pdf.

Chapter 5
Meta's Counter Conversations Project and Google's Search Redirect Project

Abstract This chapter examines two tech sector-led online interventions—the Counter Conversations (CC) project and the Search Redirect (SR) initiative—as case studies in platform-based P/CVE strategic communications efforts. Both projects illustrate how digital tools, algorithms, and direct messaging functionalities can be harnessed to steer potentially vulnerable online users away from extremist content and toward counternarrative materials. Yet these efforts highlight significant challenges in achieving sustained and verifiable behaviour change. The CC project employs a one-to-one messaging approach on Facebook/Meta platforms, attempting to engage users identified as "at-risk" of adopting extremist beliefs. Google's SR method uses algorithmic ad placement to divert users searching for extremist content toward curated counternarrative playlists. While this approach potentially disrupts the otherwise linear path of "radicalising" information consumption, it is hindered by similar constraints on reliable measurement, user intentionality, and long-term effect sustainability. The chapter discusses how these methodologies struggle to address underlying socialisation processes that shape both online and offline identity formation. Comparing the CC and SR initiatives with offline one-to-one interventions highlights that while scaled and data-driven, online tools remain limited by the complexity of radicalisation as a social phenomenon. The chapter's analysis, grounded in the Reasoned Action Model and socialisation theory, suggests that practical digital P/CVE efforts require more integrated and longitudinal strategies, combining narrative credibility, meaningful engagement, and a nuanced grasp of contextual social drivers.

Keywords CVE—countering violent extremism · PVE—preventing violent extremism · Counter extremism · Communication · Facebook/Meta · Google · Redirect method

M. Zamir and C. Kaunert, *Strategic Communications' Role in Counter-Terrorism*, Advanced Sciences and Technologies for Security Applications,
https://doi.org/10.1007/978-3-031-89110-6_5

Introduction

This chapter will focus on the direct role, position, and considerations of tech sector participation in core P/CVE strategic communications and intervention initiatives. In line with the rationale of the RAM's behaviour change intentions, the AX campaign's key learnings shifted away from the tech sector's role directly and more toward the issues of what CN means, the ideological contexts of extremism, and how socialisation is present in pre-, during, and post-extremism phases. However, when considering the tech sector in a standalone context, key debates focus on how the role and scale of the internet/social media have transformed perceptions and practices of audience targeting and behaviour change understanding, with a significant emphasis on the strategic power of communication in an immersive, rapidly changing environment of narrative usage, abuse, and distortion through the ills of disinformation and misinformation.

This chapter will analyse two specific initiatives within the P/CVE strategic communications sphere that utilise tech sector platforms or tools as their primary medium of engagement. In doing so, the focus of analysis shifts the model slightly. It considers behaviour change needs in light of more detached interventions, which have relied on platform access in markedly different ways to AX (previous chapter), namely the ability to engage on a one-to-one basis via direct messaging and the use of algorithms and search engine capability to drive content to online users. One-to-one online interaction technologies will be evaluated via the CC programme utilising Facebook/Meta. Using search engines and algorithms to lead confident consumers to specific material will be examined in light of Google's SR technique. An urgent concern that must be noted now is a lack of publicly accessible information and data for both efforts.

This is a problem with P/CVE efforts in general. Still, it is compounded by the inherent 'mystery' and the tech industry's reticence to explain the mechanics and processes underlying most of its public-facing activity [17, 19]. Due to one researcher's access to the AX campaign as a practitioner, the previous chapter could be given in considerable depth, and the reduction in the current chapter's depth requires a slight adjustment in the analytical direction. As such, the main thrust of this chapter will be to explore how these two 'pilot' initiatives compare to offline efforts within P/CVE, alongside a discussion of the role online tech can play within this sphere of communication. Key elements that merge these online projects with methods of offline P/CVE are the use of strategic communications processes to reduce the risk to extremism. These diversionary pathways seek to redirect individuals away from extremist engagement/content and the use of 'credible' messengers or messages. This merging of approaches will still require a conceptual underpinning, made possible by assessing these efforts through the theory framework's model used to answer the central questions of this research project. For this chapter, this model can be applied in the following manner:

RAM: How platform-based 'detached' interventions seek to alter and modify the behaviour of 'at-risk' groups/individuals.

Socialisation Theory: What social engagement and normative alignment processes are relevant to such interventions concerning addressing extremist ideas and seeking to divert users away from them? RAM and Socialisation theory merge into the SoRaSCo model but with more emphasis on the informal role of tech as a socialising agent.

Tech-Based Communication: How does tech support strategic communications aims?

Both initiatives will be reviewed using the RAM and socialisation theory frameworks and the aforementioned strategic communications considerations. The focus will then shift to a comparative analysis of these efforts with offline efforts in one-to-one interventions that operate in the same P/CVE space. In doing so, the chapter is intended to provide up-to-date and relevant analysis and comparative debate to expand the efficacy and sustainability of future tech-centric P/CVE operations.

Background to the Campaigns

Firstly, the CC programme will be examined, followed by SR. Both case studies will be detailed using their respective concepts, methodologies, outcomes, and offline comparisons. They will also critically examine where these issues leave the tech and P/CVE sectors in an evolving global security and social change environment. The conceptual frameworks will be valuable conduits for conducting this element, with specific and broader variables brought into play. At the heart of this chapter, the suitability and efficacy of radicalisation will be a crucial feature. This is in line with the overall emerging findings of this research project, namely that the concept of radicalisation creates unrealistic and methodological weaknesses in P/CVE efforts that, as a result, affect the role of strategic communications efforts.

Case Study 2: The Counter Conversations Project (CC)

The initial One2One pilot project will be referred to as Counter Conversations or CC for the sake of clarity and to assist the combined analysis of the phase two iteration of the original. CC began as a pilot initiative that cooperated between the tech industry (Facebook), civic society (ISD), and Australia's Curtin University. In this respect, Facebook's official involvement was that of a collaborator, not a creator, given that ISD provided access to its Against Violent Extremism (AVE) network of 'former extremists' and Curtin University's academic credentials for measurement and assessment purposes. Due to the lack of publicly available information on CC and the AVE network, it is essential to define the project's stated objectives, methodologies, and processes using the available sources at the minimum. This should make it possible to offer discourse and critical analysis based on lessons learned and assessments of comparable work to arrive at significant conceptual and practical issues for future efforts in this field. Therefore, the case study review of CC begins with an outline of the project's overall aims and processes, assessing any parallels with offline efforts and rationale-based arguments as to the approach's merits. This discussion will be permeated with an analysis of how the theory frameworks can help highlight essential learning elements where relevant.

Design Considerations

In its first 'pilot' iteration, the project stakeholders emphasised the remit of seeking to test the viability of such an approach, given it was a move away from offline efforts within P/CVE towards accessing and harnessing tech tools, platforms, and processes to deliver core counter-extremism work. The approach was to create a direct online one-to-one intervention using instant messaging as the communication medium. In this regard, it appears Facebook's direct messaging utility best served project needs as it was easy to access, linked to one of the biggest social engagement platforms in the world, and relatively cost-effective for a pilot project's needs. A core aim of the project was to assess the effects of direct messaging on online users displaying or expressing extremist ideas and narratives. The underlying objective behind this aim was to dissuade such individuals from going further down this path via this intervention. At a technical level with the CT lexicon of practice and learning, this effort, therefore, sits in the 'prevention' sphere, which is based on seeking to communicatively prevent individuals from traversing further down the road of extremism towards violence.

Much of the project's written information suggests that using Facebook was also predicated on a little-known utility known as 'pay-to-message'. This facility allowed for someone to pay a nominal fee ($1) to be able to direct message a person regardless of whether they were 'friends' on Facebook or followed each other. Using another tool known as 'Graph Search' that at the time allowed users to search Facebook by demographic factors such as age, location, and relationship status, in addition to interests, pages liked, and group membership, enabled the project team to target their searches using these factors to their advantage. This appears to have been a significant advantage on identifying online users through various search filters and, to some extent, limiting speculative engagements to the minimum. It also permitted in-depth messaging and communication options, which offer the potential for wide-ranging discussions and negate word limits and other restrictions often associated with social media platforms. Therefore, while it is commonly accepted in research circles that Twitter (X) is frequently the go-to platform for researchers seeking to map extremist ecosystems, Facebook appears to have been deemed to be more appropriate when attempting to identify at-risk individuals within specific demographic criteria and matching those individuals to intervention providers.

From a methodology perspective, the choice of platform decision was made to allow for access to users and content that was more personalised than generic. Added to geo-location data and age, gender, page 'likes', and follows, this approach provided a mini 'individual eco-system' for the team to tap into. The matching of intervention providers according to location, age and gender data further negated (within the context of online engagement) the idea that such 'cold-calling' direct messaging could be mismatched to the target recipient.

The use of intervention providers/formers is not, at face value, an overly controversial practice within prevention efforts in general when looking at gang prevention work, harm reduction efforts, etc. Even in the CT context, if the UK's prevent strategy is used as an example in this regard, the entire approach of the 'Channel' programme is built upon the use of 'specialist' intervention providers, who engage and seek to develop a risk-reduction approach through direct engagement with 'radicalised' or vulnerable individuals [6–8]. CC used the same methodology via accessing the AVE network, which is a loose-knit affiliate-based global group whose main entry requirement for membership is having some form of personal contact with/background in extremism, extremist groups, or violence. This network is, to a lesser or greater extent, the central entity that has amplified and normalised the term 'former' to denote those people who have had direct experience of extremism. Being known as a 'former' carries a dual meaning in that it presupposes a type of experience that is unique but also suggests a form of charisma in the realm of credibility regarding knowledge, awareness, and understanding of extremism.

From a theory framework perspective, such intervention providers become the critical tool in the RAM's behavioural control phase, but through direct one-to-one communication and not the model's traditional usage of mass communications in behaviour change efforts. This point of nuance then sends the behavioural control phase inversely towards the model's perceived norms and normative control stages. In other words, the intervention provider becomes an agent of socialisation seeking to re-socialise the individual through normative realignment and subsequent alteration of behavioural attitude and intention. This issue will be critically analysed alongside previously mentioned considerations in more detail once both case studies have been outlined.

As the original project aspired, in part, to bring offline intervention learnings to the online realm, publicly available information on the pilot suggests that all intervention providers hired to participate in this project had some form of offline intervention experience. It is unclear how robust this prior experience was, under what auspices it transpired, and what differences there may have been in regional/national interpretations and approaches to terms like intervention, vulnerable, radical, and extreme. Ten former extremists from the AVE network were involved, five former Far-Right extremists from North America and five former Islamist extremists from the United Kingdom. This grouping exemplifies the previous point about variational differences being a possible extraneous variable within this delivery phase. One woman and four men comprised each ideological group.

The pilot phase report suggests this group allowed the team to test the methodology across regions, ideologies, and gender. A small remuneration of £100 per month was paid to these intervention providers, who were acting nearly voluntarily. Other features of this process included offering intervention providers the option to remain anonymous and being given new Facebook accounts created by the project team specifically for this project. The project team monitored these accounts to code messages and responses, ensure a timely response to more responsive candidates, and protect both intervention providers and candidates if an urgent threat arose concerning potential risk of harm or recruitment. An area that requires further research and data is the ability to compare offline interventions and online versions and possible key variables created due to the nature and context of engagement.

After experienced intervention providers consented to participate, targeting criteria for 'at-risk' individuals had to be determined. The project team created a list of Facebook Graph Search's many search criteria and shared it with intervention providers. Instead of having the project team define the risk criteria, the intervention providers were allowed to develop the first draft. Pages were liked, and group membership was the highest risk factor. A user's cover photo and frequent posts 'tone' and 'substance' were also indicators. A possible weakness of this approach would be the significant chance of subjective interpretation of such information and any inherent bias this may create. The project team used outreach provider risk indicators to generate two sets of search criteria: one for persons at risk of violent Far-Right extremism in North America and one for violent Islamism in the UK.

External researchers hired for this project used Graph Search to identify at-risk applicants after signing non-disclosure agreements and training. The project team and intervention providers initially examined candidates to ensure they were in danger. At this point, two groups were removed: those who had been falsely identified as at-risk because they followed pages and accounts that were politically or religiously extreme but did not advocate violence and those who were so far into the extremist milieu that they were beyond the scope of this project. A potential scenario is that an intervention provider declined to contact their candidate because they thought it would endanger their safety. The often unpredictable and volatile extremist ecosystem on Facebook and social media meant that the original targeting criteria served as a starting point and developed during the study, with pages and groups taken down and others found to be shared among people at risk also added. According to the project review, the team employed a four-step approach to prepare the CC delivery for implementation. These stages were: Graph Search identifies a lengthy list of prospective candidates, researchers eliminate false positives, the project team validates the researchers' candidates, and intervention providers confirm candidates' risk.

This project's essential and critical component was the messaging process itself, from message type, content, narrative, tone, and context perspectives. It is telling that during the period of this pilot project and, to a greater or lesser extent, there remains a significant gap in the P/CVE field concerning a robust, regenerative and reliable framework for what 'good' counter-messaging or narrative should consist of. Several variations of this issue have been put forward in the last decade through network entities such as RAN and Hedayah (US State Department-funded think tank), to name two. However, accurate measurement of success criteria and replication issues mean that the 'ideal' message or CN is still a goal to work towards at the time of writing.

This research project initially offered a new conceptual framework for online P/CVE within strategic communications. In doing so, it was hoped that the dialogue of knowledge and understanding of this subject area would move further along. This is an issue for the concluding chapter regarding how successfully the aim has reached this target. Still, the messaging considerations for the CC pilot need discussion and analysis. As mentioned, various competing approaches exist as to what practical messaging looks like in the P/CVE space. This ranges from language usage (harm-reduction lexicons to anti-grooming and reintegration efforts) to ideological considerations and psychology-oriented efforts.

Within such a diverse opinion field, the CC approach allowed each intervention provider to build their message eco-system based on their preferences, experiences and communication style. This does, of course, mean that the overall messaging approach comes across as slightly ad-hoc and lacking methodological rigour. Still, equally, the approach also lends itself well to testing and trying out different engagement tactics to build out a bigger picture of the iterative intervention process via targeted communications. In many respects, therefore, this approach has become symptomatic of P/CVE efforts online and offline over the past decade, namely pilot project after pilot project, but no official, sustainable, and measurable core method of approach.

Of equal importance to this issue is that such approaches cannot, by default, build, bridge, or merge inter-disciplinary methods from within persuasion, narrative and strategic communications practices on the one hand, alongside cognitive-behavioural needs and methods of engagement on the other. This is potentially why using the combined RAM alongside socialisation theory and strategic communications methods offers more than just a framework to review such efforts. Moreover, this theory framework provides insight into why certain online P/CVE efforts generate more questions for an already convoluted discourse and why these questions continue to go unanswered. The short answer to this difficulty is that no two initiatives have any core thematic or conceptual alignment beyond being P/CVE efforts. One of the reasons for this is the insistence that online P/CVE efforts quickly and miraculously offer scalability and vanity metrics as measures of success. This is something chapter four has already highlighted in detail, but suffice to say, this is both a policy and practice weakness currently still not being taken seriously.

Critics of this argument will point to differences in target audiences, demographics, cultural nuance, etc., as being reasons why overt replication of such efforts is not feasible, but as will be demonstrated when offline P/CVE and prevention approaches are reviewed, this is a weak argument for one simple reason; it is precisely because audience nuance, theological/ideological issues are absent from strategic-level understanding and replaced with an over-reliance on mass communications and psychology tools, that so many P/CVE efforts cannot prove any tangible, replicable, and sustainable form of behavioural or prevention success. This overall argument and the usefulness of the theory framework applied to this research effort will be given due attention in the final chapter of this work. Still, these themes cannot be ignored within the critical analysis or review of case studies within the online P/CVE context.

Socialisation Theory Considerations Within the Project's Parameters

The CC project seemed keen on being an iterative effort, observing intervention providers from different walks of life devising their unique approaches to engaging with different 'levels of risk'. This point is suggestive of a broader debate within P/CVE circles, namely that the most effective and, to some extent, most observable form of intervention is an individual-level engagement between a 'messenger' (person or narrative) that possesses a specific skill-set and contextual credibility level and someone (recipient) either already involved with extremism or who has come into contact with extremist ideas. Extremist ideologies morph and evolve into subtle variations of their original iterations but do not radically alter their core narrative and worldview or, as Max Weber championed, 'weltanschauung' (Best, 2003). In terms of tackling Jihadi ideology, it can be argued that the P/CVE worldview has been too skewed towards a Western understanding of the problem set and root causes. In contrast, the extremist narratives themselves are adept at being contextually relevant and providing the exact nuance that audiences seek out to help their social realities. This can be seen as both a problem for intervention and an opportunity to improve design and narrative issues. This goes to the heart of one of the theory frameworks this project has used: socialisation theory.

CC Project Outcomes Applied to the RAM

The main findings of the original pilot of CC yielded some exciting outcomes but perhaps nothing overly revolutionary in terms of existing knowledge within P/CVE. A worthy point to note here is that given how these messages were intended to be delivered; the project team seem to have taken gratification from the percentage of messages that were 'received' by the final confirmed 'at-risk' cohort. There is a simple explanation for this sense of satisfaction when considering how low recipients read the percentages of unsolicited 'direct-mail' or messages. 60% of messages sent by intervention providers were received with no discernible difference between profiles of the Far-Right and Islamist-centric accounts. This read rate is considerably higher than average for unsolicited email campaigns. MailChimp, a provider of email marketing services, reports that no industry has an average read rate that exceeds 30%. 59% of the audience who received these messages responded in some way, directly or indirectly, demonstrating a change in behaviour, such as altering their privacy settings or blocking the provider [3].

Without an actual direct 'reply', these shifts in behaviour were coded as 'reactions'. There was a significant difference in response rates between Far-Right and Jihadi candidates, with 63% of Far-Right candidates responding compared to 42% of Jihadi/Islamists. Again, compared to typical email marketing campaigns, these response rates are orders of magnitude higher than expected. No industry tracked by

MailChimp achieves click rates (indicating the recipient engages with the email's content) above 6%. Legal frameworks in the United Kingdom and the United States could be a factor in the disparate response rates between groups. In the United States, candidates openly support racist ideologies and pose with firearms. This may indicate that US candidates are more likely to believe they are immune from prosecution under the First and Second Amendments and, as a result, are more likely to engage in political discourse. In addition, the media coverage of Daesh and associated groups during this pilot project may have prompted candidates who expressed sympathy for Daesh to exercise greater caution.

If RAM is applied to this effort, the processes within the model itself become confusing for this example. The simple reason is that the only real contextual and background factors made available in the literature are that participants were vulnerable or 'at-risk' from extremism and possibly violent extremism/terrorism. It is unclear which part of the RAM process (see diagram below) the intervention has occurred. This means that tackling behavioural change in this context would only be speculative (Fig. 5.1).

Since this was a pilot initiative (as mentioned earlier in this chapter), there was no evaluation of and agreement upon 'best practice' for the message type for such a unique target population. Offline interventions have differing views on tone and ideological dialogue. The project team chose not to tell intervention providers what to say. As previously discussed, each practitioner could create messages based on their

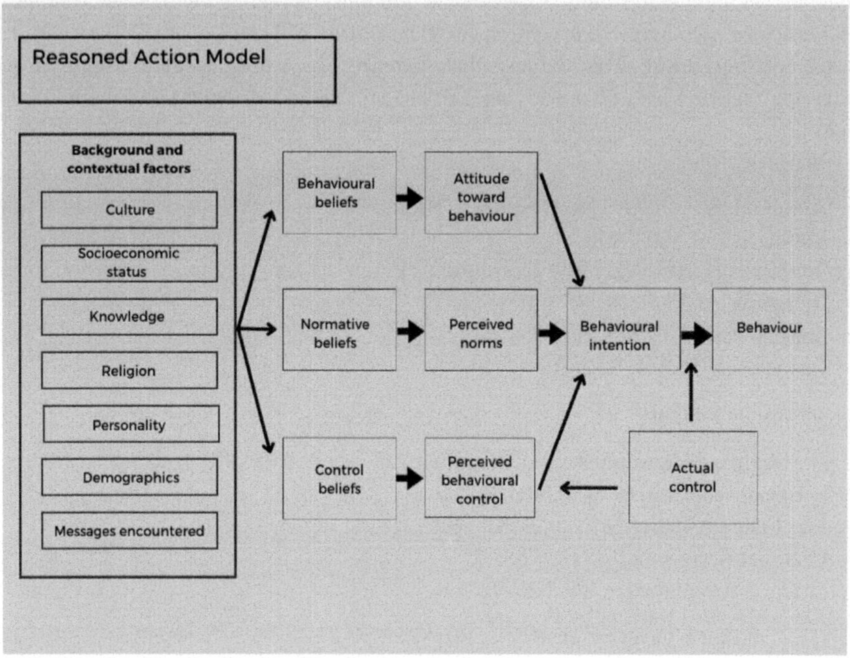

Fig. 5.1 Reasoned action model

intervention expertise and personal background. This appears to have allowed the project team to gather data on the message types most likely to elicit a response and bring much-needed data to an ongoing conversation among intervention providers. When considering response types, some candidates denied their beliefs or refused to participate. However, most were willing to talk to outreach providers to some degree. In this context, the team settled on five or more texts and 'meaningful' dialogue with the intervention provider was considered sustained involvement. The term meaningful is open to huge degrees of interpretation. Some of the key data in this regard are below:

- 12% = Denied Involvement
- 8% = Brief Engagement
- 20% = Non-Engagement
- 60% = Sustained Engagement

One intervention provider had a ninety per cent (90%) response rate, while two had none/zero. It appears to confirm another standard P/CVE industry narrative: many offline intervention providers have an unquantifiable 'x factor' that others do not. The project team coded communications by tone and content to find commonalities in successful messages. This is claimed to have helped them 'develop data-driven online' intervention best practices. The team's findings suggest that message tone is crucial, with some generating zero reactions and others, like casual and sympathetic/understanding, eliciting over 80%. "Hey, you're an MMA fighter?" is a casual message. "I understand" and "I know what you must be feeling" were seen as honest and emotive messages. Intervention providers often reflected on their past and its impact on their future. This suggests that changing tone can significantly improve an intervention provider's response rate. Data to provide more context to this finding is below.

Message Tone:

- Antagonistic = 100% No Response (NR)
- Meditative = 100% NR
- Scholarly = 70% NR, 30% Response (R)
- Reflective = 25% NR, 75% R
- Sentimental = 17% NR, 83% R
- Casual = 17% NR, 83% R

Message Content:

- Highlighting Consequences of Negative Actions = 85% NR, 15% R
- Personal Question = 64% NR, 36% R
- Ideological Challenge = 58% NR, 42% R
- Personal Story = 22% NR, 78% R
- Offer of Assistance = 8% NR, 92% R

([3]/Counter Conversations/Meta, 2018).

Tone and content were therefore measured with keen interest. Results show patterns and best practices according to the project team, e.g., that personal stories

It appears that the CC communication and engagement tactics did contain an ability to deliver effective messaging via CN auspices using 'credible' intervention providers. However, in risk assessment, strategic intent, and creating behaviour change processes, the tactic does not seem well-developed enough to effectively counter pre-existing socialisation issues, which then become inhibitors to achieving behavioural control from the intervention providers' perspective rather than the recipient's. In other words, the recipient retains too much control over the intervention, both motivationally and normatively. This leads to the issue of the power of socialisation processes being overly dominant variables in such an engagement. In such a case, behaviour change and strategic intent to achieve become hopeful, not core elements.

Learnings and Further Considerations of the CC Project

Socialisation theory posits that people are socialised into moderate values and extreme ones. The process is the same, while the intentions and outcomes are different. As a pilot project, which later was extended to a bigger scale and with more resources and online tools, CC went some way to replicating the offline intervention sphere. Still, overall, the data and outcomes could be more in-depth and revealing enough to make better quality assertions regarding online communications within the P/CVE context. Regarding seeking to shift behaviour, the CC experiment needed to be longer to measure long-term belief systems or behaviour changes. Several interactions suggested persistent involvement and trust between intervention providers and candidates could lead to long-term behaviour change. Some applicants said they trusted their intervention provider and that it made them 'think deeply' for the first time. Other conversations showed an openness to new ideas and the intervention providers' journeys. One candidate said, "I am curious to hear how and why you've changed your mindset". "I don't need no help", the candidate said during another intervention, "I like myself". The intervention ended without further pressuring the candidate (One2One, 2016; CC, 2018). One week later, the same candidate told the intervention provider, "I need help if u can help me".

This emphasises the necessity of communicating with at-risk individuals and the long-term effects of this approach. Long-term and sustained engagement may offer the best chance of monitoring and achieving behaviour change. Google's SR may provide some more developmental knowledge to add to this ongoing effort within P/CVE to use tech as a critical tool to fight hate and extremism online. Before SR can be discussed, it is essential to contrast the pilot CC effort with its more large-scale second iteration in 2017. The main adaptions in the second iteration were that machine learning was used via Natural Language Processing (NLP) to attempt to create auto-generated identification of potential candidates, the intervention provider network was diversified to include survivors and professional counsellors, and a more robust risk-assessment methodology was developed.

Again, the data is generally anecdotal and does not go into sufficient depth to create a core hypothesis regarding what works and what does not work in this space. One example should be illustrative enough of this point. Professional counsellors had more 'conversations' than former extremist intervention providers did, but the latter was better at sustaining conversations. It is hard to make any robust assertion from such data when both outcomes appear to be given a form of interpretation that simultaneously makes them appear similar yet suggestive of significant variations. The primary learning from this second phase is something P/CVE has become well known for; most projects that continue to develop do so with a greater emphasis on adding mainstream elements instead of bespoke elements. In other words, they become more generic and universal rather than more streamlined and nuanced. This may be the case with Google's SR method, which will be discussed next.

Case Study 3: Search Redirect Project (SR)

The Redirect Method, or SR project, has garnered considerable news interest as a potentially successful strategy for reducing the availability of internet information that enhances the attractiveness of VSOs, encourages radicalisation, and may make viewers more vulnerable to recruitment [14, 18]. In essence, this initiative was a tech/civil society/think tank partnership between Jigsaw (a subsidiary company of Google), the NextGen foundation and, in some key phases, Moonshot CVE. As previously introduced in the AX case study (Chap. 5), SR's strategy was also to use Google 'AdWord' tech as a tactic/tool to detect those searching for violent extremist content on Google and then expose those individuals to advertising that links to CN communications in their search results. As a tactic to try and divert the individual's attention away from extremist content, this does appear even at the outset as a hit-and-miss approach, given that YouTube ads are short, often easily bypassed and forgettable. SR was initially tested between August 2015 and March 2016. The researchers found content aimed at 'debunking' Daesh narratives among prospective sympathisers. The most engaging videos were collected and shown in playlists explicitly created for the initial dissemination phase. These playlists were sent worldwide to those whose search habits indicated a favourable attitude toward Daesh.

Background Context to SR

Interestingly, the reliance on the term 'attitude' links the assertion neatly with the RAM's behavioural attitude and intention phases, respectively. However, it falls short of being a fundamental reason for offering an individual CN content. The reason for this is that it is impossible to predict or to know if engaging with online extremist content is enough of a reason to suggest a person has adopted extremist views. The same person's level and level of nuance in their layers of socialisation may already contain protective factors to limit any potential harmful effects of such content. Another example of how vanity metrics are used in proposing 'success' criteria (Chap. 4) is seen in the outcome data presented during this initial phase of the rollout of SR. These initiatives are said to have 'reached' an estimated 320,906 unique individuals over eight weeks by exposing them to a total of 500,070 min of material (Redirect Method, 2017). Chapter 4's discussion on vanity metrics outlines some key elements as to why these metrics need to be taken with caution when looking at specific effectiveness indicators in a strategic communications context. In short, they do not really imply or point to any tangible evidence or correlation to overt behaviour change.

SR's Objectives

The purpose of SR seems to have been (even in its embryonic phase of deployment) to tailor this tool to groups in the US, targeting not just Jihadi violence but also Far-Right extremist violence, particularly from Neo-Nazi and White Supremacist movements. Despite the high number of attacks committed by Far-Right extremists, P/CVE efforts and global governmental funding, in general, have previously paid little coordinated and sustained attention to the violent Far-Right, aside from some programmes in the European context. More strategically-driven efforts in Germany, the UK, and Denmark have been exceptions to this pattern in the last five to six years. Given the size and scale of the prevalent issues within the US context concerning the burgeoning threat from right-wing extremism, this approach is needed [16].

Jigsaw and Moonshot CVE's implementation of SR presents the issue of how many online P/CVE strategic communications/prevention campaigns have very short lifespans and then undergo many changes before being promoted again. Whilst this approach is not overly hazardous, nuance and detailed findings often suffer due to short-termism and parameter flux concerning content design. Despite impressive initial figures, citing 320,000 online views, watching a combined 500,00 min of 'redirected content' [14], the issue with SR seems to be a lack of reliability regarding longer-term effects or usage. Approach-wise, one arm of the dissemination targeted violent Islamist/Jihadi extremists, while the other arm targeted violent Far-Right extremists. The primary campaign targeted 50 states in the US and lasted four months for Jihadis and three months for Far-Right extremist elements. Moonshot CVE then identified ten counties at the state level whose users conduct the most significant number of extremist searches per capita for each of the two extremist groups and subsequently concentrated a month-long campaign on each of these states [13]. It would be fair to say that this initiative could benefit from at least a medium-term delivery approach, which generates richer and more relevant data. This leads to two main concerns with approaches such as the SR.

Design and Delivery Issues for SR

According to tech observer Kieron O'Hara, there are two key concerns. The first is that success in the present with some 'innovative' new approach does not always correlate to success in the future when all actors have acclimatised to the new environment. Regarding behavioural change contexts, this issue can work for and against a tool like SR. When online users search, the underlying assumption is that the results are the outcome of a 'neutral' algorithm (controversy around algorithms' inherent bias is an ongoing concern). Still, the advertisements are placed by individuals who want users to click on them for their commercial, religious, or ideological goals. SR contradicts this notion: the ad is constructed and displayed for the user's benefit, not the advertiser's [14]. It aims to control and, therefore, modify behaviour (according to the advertiser's concept of more appropriate behaviour). In this guise, SR seeks to exercise behavioural control according to the RAM approach. The issue appears to be that as the CN content is only in the form of an 'ad', users can alter their behaviour pattern and develop a new set of normative practices (perceived social norms for online engagement with extremist content), which could nullify the strategic intent of the SR content. The AX project had the same issue with its use of targeted ads in that, in reality, they are only 'targeted' if the user clicks on them. As such, users intentionally seeking to source and engage with such content can re-socialise their online practices to suit their behavioural intentions with little risk of an external behavioural control ruining the process.

As knowledge and awareness (or socialisation processes) of extremist online behaviour develop, the risk is that an approach like SR eventually begins to undermine its objectives. This is because what it sets out to achieve can be offset negatively by users adapting their behaviours to avoid SR tools. Additionally, individuals who seek the allure of identity, a sense of belonging, and loyalty to a cause via extremism are searching for something more innate and powerful. While diversionary tactics within P/CVE are necessary, SR will not help such 'seekers' achieve their objectives. They will continue to look and parse information according to the specific intent, motivation, and behavioural needs of those seeking something more visceral than an 'ad'. In defence of SR, Jigsaw/Moonshot CVE researchers are aware of this issue and plan to integrate therapeutic tools such as online therapy in subsequent phases. This additional layer of detail and context would greatly benefit this approach [14]. In essence, this suggests there is an acknowledgement that approaches such as SR need a more holistic set of parameters and evaluative metrics to be effective in the long run. These could include cognitive, behavioural, and dissonance controls and tiered content with an incremental approach to the type and tone of narrative required at different stages of socialisation into and out of extremism [2].

Should such innovations and enhancements emerge, the more cerebral question is around whether or not a tool like SR can positively affect the behavioural action phase of the RAM model with a behavioural control intervention. Could this occur without compromising other essential components of our society, such as free speech and information access? The fact that such tech is relatively private and reproducible in different guises means that good and evil 'actors' can enjoy the tool for its communicative and persuasive qualities. This creates a plethora of questions about the motives, means, and use of data, privacy and moral narratives in and around such processes.

Moreover, much of the work is premised on the ability to track and analyse online users' search behaviours. The SR method mostly appears to affect content page links from search results in this regard. Even though no one accuses Jigsaw of doing so, it would be easy to use the methods to censor the content searched for (it is already common practice to do so for 'child porn', and Google is similarly attempting to suppress 'revenge porn') or to classify the searcher as a potential terrorist based on highly uncertain evidence that will result in many false positives [14]. Where would this method draw the line regarding what it seeks to redirect, for whom, and to what end? Search results or criteria alone cannot feasibly identify motive, attitude, and behavioural intention. Other facets of the socialisation process into extremism are of high relevance to correctly identifying risk or vulnerability in this context. This point leads to concerns regarding how much SR can be effective in the long term. The accuracy of results regarding being adept at filtering harmful from non-harmful content is still an art and not a science (despite the A.I. factor).

Whilst an enormously popular search engine, extremists can and do seek alternative means to Google, to teach their narratives to followers and 'seekers' alike. Encryption-based messages and platforms are but one example. As mentioned earlier, there is a high level of need for the user to be the actual architect of his/her behavioural control by clicking on the ad. This high-risk approach can be circumvented over time by these online users by becoming socialised in new normative practices for online behaviour. These issues bring forward moral and ethical concerns around limiting free speech and expression, with questions around transparency and bias also being relevant.

As many have argued, we are living in a post-truth world. This is characterised by sentiment overpowering facts, where falsehoods and truth compete on an equal footing on the internet and where faith in authority and conventional gatekeepers of material is at an all-time low [9]. Many consider scholarship, science, evidence, peer review, experience, and competence not as trustworthy verifiers of the truth but as irrefutable evidence of an establishment conspiracy and seek to express power narratives dressed as popular social movements on the internet [15]. The use of and ownership of narratives in the online domain appears to be as much about understanding where 'power' lies and where mainstream values do or do not reside as a measurement of social order [1, 5, 9, 15]. SR is an ingenious technological project that helps deflect people away from genuinely abhorrent and terrible apocalyptic nihilism, but it does little to restore the legitimacy of mainstream viewpoints. In this regard, it is possibly an example of how established authorities have adapted to the post-truth era. By tacitly placing themselves alongside extremist actors like Daesh and posing themselves as equal (though better resourced) 'players', might such authorities perhaps undercut their own beliefs of uniqueness and superiority? This suggests that P/CVE CNs operate almost as counters to existing counters and

Analysis Using 'SoRaSCo' Model - Search Redirect Case Study

CRITERIA	SEARCH REDIRECT PROJECT TACTIC	SOCIALIZATION CONTEXT	COMMUNICATIONS CONTEXT
BEHAVIOURAL INTENTION	USE OF CONTENT REDIRECTION TO AFFECT INTENTIONS	SEEK TO CHANGE INTENTION AND ONLINE ENGAGEMENT VIA REDIRECTING USER	TARGETED ADS DISPLAYED TO USER
ATTITUDE TOWARDS BEHAVIOUR	USE OF CONTENT REDIRECTION TO AFFECT NORMS IN POSITIVE MANNER	SEEK TO CHANGE ATTITUDE VIA ONLINE ENGAGEMENT/ REDIRECTING USER	TARGETED ADS DISPLAYED TO USER
PERCEIVED/ SUBJECTIVE NORMS	HOPE PARTICIPANT BEGINS TO QUESTION EXISTING NORMS VIA NEW CONTENT	SEEK TO ALTER NORMS VIA ONLINE REDIRECTION OF CONTENT PREFERENCES	TARGETED ADS DISPLAYED TO USER
PERCEIVED BEHAVIOURAL/ ACTUAL CONTROL	HOPE NEW REDIRECTED CONTENT CONTROLS KEY FACTORS AROUND BEHAVIOUR	HOPE RESOCIALIZATION OCCURS VIA ONLINE CONTENT REDIRECTION	TARGETED ADS DISPLAYED TO USER
BACKGROUND CONTEXTUAL FACTORS	REMAIN AS 'KNOWN UNKNOWNS'	HOPE THESE DO NOT NEGATIVELY AFFECT REDIRECT PROCESS	TARGETED ADS DISPLAYED TO USER
FEEDBACK & ITERATIVE APPROACH	USE OF VANITY METRICS AND DATA ANAYSIS	ADAPT REDIRECT PROCESS AND CONTENT TO IMPROVE EFFECT	USE DATA TO IMPROVE TARGETING/CONTENT REDIRECT ELEMENT

Fig. 5.3 Search Redirect and SoRaSCo model analysis

not as representations of an established order or master narrative per se (Fukayama, 2018). Where possible, these seemingly convoluted and disparate elements require analysis of the project's known auspices using the SoRaSCo model. In doing so, areas of strategy, behaviour change, and re-socialising elements may be identified. Analysis of SR using the SoRaSCo model is below (Fig. 5.3).

RAM Approach and Socialisation Analysis of SR

In terms of the RAM approach and socialisation theory, this objective is too generic and lacks the sophistication of understanding and knowing the actual needs of the intended target audience. The RAM is a theory of human behaviour that suggests individuals make decisions based on their intentions, which are formed by their attitudes and subjective norms. It is impossible to infer intentions or possible types of behaviour from a few video views. Equally, socialisation seeks to explain how individuals learn and adopt their society or culture's norms, values, and expectations. Again, seeking to change or alter behaviour through SR contains too many extraneous variables regarding how many different social, cultural, and ideological factors may be relevant for one group over another. The approach seeks to offer nuanced and relevant content, but what parameters is this nuance based on? This means that to effectively address the causes, manifestations, and nuances of extremism, P/CVE needs to have a whole-society approach and not merely a communication, counter-extremism, or security approach. The cognitive cannot change without the social. Therefore, the narrative can only be changed if the social/environmental issues and motivation/intention issues change alongside it.

Strategic Communications Perspectives as Applied to SR

From a strategic communications framework perspective, the objectives of SR are bold in terms of how they read. Available source material points to a significant objective: 'preventing unfettered access to extremist information'. What is striking about this from an effects outcome and design perspective is that the online eco-system alongside the individual's eco-system of engagement, normative alignment, and acceptance/rejection of ideas and narratives are enormous chasms of space. An online intervention not linked to offline or other support/signposting processes is merely embedding itself in a small corner of the cognitive and social reality of the user. Both reasoned action and socialisation theories provide ample context to the intricate web of influence and communication elements in any individual journey. One communications-based engagement (SR) is merely added to the plethora of others competing to lead the individual to what lies beyond behavioural intention in the TRA model, which is action.

This is also relevant to SR's objective: 'to pierce an uninterrupted radicalisation path'. The literature review of this research project discusses, in much detail, the issue with the term radicalisation, its apparent 'process' and 'flow', alongside the problem that the concept is either too broad in some of its uses or too narrow in other guises. It is also problematic because its starting point (onset of radicalisation) is just another layer of socialisation and should be treated as such. A third objective also appears, which suggests the presentation of counter-radicalisation to viewers is in and of itself a vital part of the project's focus.

Key Learnings from the Analysis of SR

Using the SR method can come across as a scatter-gun type approach or objective, which can only be equally vague and inconclusive in terms of yield. The main issue with SR from a strategic communications perspective is that it needs to rely more on vanity metrics. This approach offers some essential elements and understanding of engagement-related outcomes, but this is far removed from being able to identify, track, and contextualise robust behaviour change analysis for the simple reason that it would be impossible to ascertain other social, emotional, narrative, and cognitive elements that could contribute to the rejection of extremism. One of the key findings of this entire research effort is that P/CVE CNs will never really be able to prove they alone have countered the extremist narrative and the affiliation, sense of belonging, and loyalty that such narratives offer.

Whilst still somewhat premature in the conclusion, this appears to be a thread in all the case studies and the theory frameworks used in this study. This can be seen from a cynical perspective as P/CVE efforts attempting to look busy preventing violent extremism but not being able to evidence it, or a more measured proposition that acknowledges P/CVE efforts as trying to play some role in a menu along with other interventions (formal and informal, known and unknown) that all contribute to reducing the threat from violent extremism. What is emerging at this stage in the context of a more concrete finding, however, is that SR and, to some extent, CC are seeking prevention through a diversionary pathway approach. These initiatives appear to be relying on a mixture of the following inputs.

There is an emphasis on using communication and tech sector mediums and propriety knowledge as tools created with a specific strategic intent and a desired end-state objective. This suggests such projects are designed with overt aims and objectives in tow. Another key feature is using 'credible' messengers (formers) or messages (CNs). There is a prominent element of relying on the concept of radicalisation to understand the cause-and-effect parameters relevant to the type of intervention needed. This reliance contains many assumptions that remain untested and empirically weak. Using narratives to counter existing narratives suggests two things: firstly, that extremist narratives are fully understood (nuance included), motivational features of why people extremist groups are fully understood, and the specific components of what makes a counter to this narrative effective are also understood. This latter point is incorrect because most online P/CVE interventions are continually known as pilots, test phases or alike. This points to the fact that no overarching and robust framework is in place (the critical aim of this research project). Secondly, it is informally implied that a CN must, in essence, contain content and form of an exceptional standard to attract, engage, and cause the recipient to question to a sufficient level to lead to behavioural control, re-socialisation, and, ultimately, behaviour change. In defence of SR, this method has gone on to be used in various other P/CVE-related efforts in different parts of the world. It seems to be a tactic that is viewed in some way as cost-effective and able to deliver 'results', albeit heavily dependent on vanity metrics. The next chapter brings learning from the case studies and related contexts to provide an overarching discussion of the varied and evolving landscape of P/CVE strategic communications. In doing so, this discussion allows for a segue to some of the main findings from this research study.

References

1. Beck, U. (2009). *World at risk*. Polity.
2. European Union (2017). *Identity, Belonging and Extremism*. [online] EU Home. Available at: http://www.congress-intercultural.eu/en/initiative/215-identity--belonging-and-extremism.html. [Accessed 24 Nov. 2021].

3. Frennett, R., & Dow, M. (2015). *One to One Online Interventions—A Pilot CVE Methodology*. [online] ISD. Available at: https://www.isdglobal.org/isd-publications/one-to-one-online-interventions-a-pilot-cve-methodology/ [Accessed 20 Mar. 2022].
4. Fukuyama, F. (2019). *Identity: Contemporary identity politics and the struggle for recognition*. Profile Books.
5. Giddens, A. (1990). *The consequences of modernity*. Polity Press.
6. Hassan, A. (2012). Shifting paradigms. *Arches Quarterly*, 5(9).
7. Home Office OSCT. (2009). *Contest Strategy 2*. Home Office Office for Security and Counter Terrorism.
8. Home Office OSCT. (2011). *Prevent Strategy Review*. Home Office Office for Security and Counter Terrorism.
9. McIntyre, L. (2018). *Post-Truth*. MIT Press.
10. Meta, F. (2016a). *Counter Speech: extreme lives*. [online] Counterspeech. Available at: https://www.facebook.com/en/initiatives/p2p-facebook-global/ [Accessed 21 Mar. 2022].
11. Meta, F. (2016b). *Counter Speech: peer 2 peer*. [online] Counterspeech. Available at: https://www.facebook.com/en/initiatives/p2p-facebook-global/ [Accessed 21 Mar. 2022].
12. Meta, F. (2016c). *Counter Speech: redirect*. [online] Counterspeech. Available at: https://www.facebook.com/en/initiatives/redirect/ [Accessed 21 Mar. 2022].
13. Moonshot CVE (2021). *Canada Redirect—Final Report—Moonshot*. [online] moonshotteam.com. Available at: https://moonshotteam.com/resource/canada-redirect-final-report/ [Accessed 19 May 2022].
14. O'Hara, K. (2016). *The limits of redirection*. [online] Slate. Available at: https://slate.com/technology/2016/09/the-problem-with-google-jigsaws-anti-extremism-plan-redirect.html [Accessed 20 Jan. 2023].
15. Richards, J. (2021). Extremist propaganda and the 'Politics of the Internet'. *The Journal of Intelligence, Conflict, and Warfare*, 3(3), pp.139–141. https://doi.org/10.21810/jicw.v3i3.2798.
16. Robertson, C., Mele, C., Tavernise, S. (2018). 11 Killed in Synagogue Massacre; Suspect Charged With 29 Counts. *The New York Times*. [online] 27 Oct. Available at: https://www.nytimes.com/2018/10/27/us/active-shooter-pittsburgh-synagogue-shooting.html.
17. Schick, N. (2020). *Deepfakes*. Twelve.
18. Wired Staff and Greenberg, A. (2016). *Google's clever plan to stop aspiring ISIS recruits*. [online] WIRED. Available at: https://www.wired.com/2016/09/googles-clever-plan-stop-aspiring-isis-recruits/ [Accessed 4 May 2022].
19. Zuboff, S. (2018). *The age of surveillance capitalism: The fight for the future at the new fontier*. Basic Books.

Chapter 6
Broader Strat Comms and P/CVE Reflections

Abstract This chapter examines the strategic communications dimension of online Preventing/Countering Violent Extremism (P/CVE) initiatives. It highlights the importance of integrating clear, theory-driven design processes, culturally resonant narratives, and robust evaluation methods. Key discussions include moving beyond vanity metrics towards meaningful audience engagement and behaviour change, employing behaviour change theories (such as the Reasoned Action Model) alongside socialisation frameworks, and leveraging advanced analytic tools (including sentiment analysis) to capture subtle shifts in attitudes. The chapter underscores that achieving measurable, long-term impact requires not just tailored, credible, and adaptive messaging, but also iterative feedback loops, ethical considerations, and close integration with offline interventions. Ultimately, it argues that more nuanced, context-sensitive, and evidence-based strategic communications approaches are essential for effectively countering extremist narratives and fostering resilience in at-risk audiences.

Keywords CVE—countering violent extremism · PVE—preventing violent extremism · Counter extremism · Communication · Strategic communication

Introduction

This chapter delves into the broader analysis and assessment of strategic communications issues within online P/CVE initiatives. The primary emphasis is placed on the following analyses and discussions:

- Comprehending the design and implementation of online campaigns
- Analysing narrative components within these campaigns
- Addressing difficulties related to measurement and evaluation
- Exploring the role and utilisation of technological tools in these endeavours

The discussion considers these elements within the overarching theory framework of socialisation and communication studies, including behaviour change needs

M. Zamir and C. Kaunert, *Strategic Communications' Role in Counter-Terrorism*,
Advanced Sciences and Technologies for Security Applications,
https://doi.org/10.1007/978-3-031-89110-6_6

(SoRaSCo). As an extension to the findings of the case studies presented earlier in this book (Chaps. 4 and 5), this chapter focuses on the interface between theory, practice, policy, and strategy within P/CVE. The following elements and sub-sections will be discussed and assessed by presenting an outline to this chapter, critically focusing on online P/CVE contexts.

Understanding and Contextualising Online P/CVE Communications Efforts

This section provides contextual and foundational learning on the process and practices of P/CVE online communications-based interventions. In doing so, it sets the scene for the components needed that ultimately combine to constitute such communications campaigns taking shape. This includes discussing P/CVE policy and strategy issues to offer context and depth of understanding of how this sector has developed.

Design and Delivery of Campaigns: Tools and Tactics of Interventions

This section explores the complexities involved in designing and implementing successful internet initiatives aimed at P/CVE. It elucidates the significance of meticulously constructed messages and strategies in countering extremist ideology and mitigating the process of radicalisation. This discussion delves into the comprehensive examination of critical elements, including audience segmentation, targeted content, and the selection of suitable communication methods. The chapter emphasises the importance of actively involving populations that are vulnerable to extremist influences and establishing collaborations with reputable individuals or organisations in the field of P/CVE.

Engagement, Measurement, and Evaluation: Behaviour Change and Sentiment Analysis

This section acknowledges the difficulties associated with quantifying the effectiveness of online counter-violent extremism (P/CVE) initiatives and the role of engagement in this process. This paper examines different assessment approaches and critical performance indicators that can be utilised to evaluate the effectiveness of campaigns. Moreover, this study delves into the importance of qualitative evaluation to acquire a more profound understanding of the evolving attitudes and behaviours of the intended recipients. The chapter emphasises the iterative enhancement of tactics by utilising data-driven evaluation.

Understanding and Contextualising Online P/CVE Communications Efforts

Online P/CVE efforts operate with a form of acknowledgement that end-users are not just passive recipients of communications but also the inherent value sets contained within these narratives. As a result, P/CVE CNs function on the logical premise that they are adequate for promoting genuine, positive alternatives to extremist content. This assumption and reasoning represent the 'what' element of the online P/CVE project's beginning phase. The second component relates to the need for a broad goal or objective, which is to combat/challenge/tackle/etc., potential support for violent extremist or terrorist organisations. The requirement that interaction and engagement with CN ideas be successful and result in a lasting legacy of behaviour change is also a part of this purpose. These goals are the 'why' behind the intervention's necessity or inception. In a relatable initiative, the GAMMMA+ model has been developed as a manual for carrying out online P/CVE interventions within a broad framework model through work by the RAN Communications and Narrative Working Group of the European Commission [15]. A vital issue for some frameworks/toolkits is to build more significant and replicable sources of best practices of CN efforts using examples evaluated through meticulous peer-to-peer evaluation, critical discourse analysis, and data analysis. Given this, RAN and similar entities could expand their remits to develop more in-depth evaluative processes alongside publishing toolkits and frameworks.

Design and Delivery of Campaigns: Tools and Tactics of Interventions

There are different tactical approaches within this field. Types of CNs include alternative narrative creation, redirecting vulnerable users, engaging users on a one-to-one basis, and offering prebunking and inoculation-style content. Narrative elements refer to the various components that make up a narrative. These elements include characters, setting, plot, conflict, and resolution. The focal point of this discussion is the examination and evaluation of story components employed in online campaigns related to P/CVE. This study explores the significance of storytelling in influencing views and refuting extremist narratives. This section provides a comprehensive primer for the following analysis in the case studies of how narratives centred around empathy, optimism, and personal growth can serve as potent instruments in deterring individuals from embracing extremist ideologies.

The fact that online identity and its compound effects on many of the audiences targeted by both extremists and P/CVE practitioners is obscured by countless and variant forms of impact criteria that cannot be documented or defined makes it challenging to isolate and verify genuine causal ties to certain decisions, acts, or behaviours. In short, it is hard to tell the root cause of the manifest effect when it comes to extremism. It is, therefore, crucial that the strategic framework developed for an online intervention offers an exceptional level of understanding of the problem set, insight into audience types and lifestyles, and critical expertise in the sub-element being explored (whether deradicalisation or capacity/resilience building). This is even more nuanced in P/CVE when theological, ideological, gender and emotional intelligence factors are present in disparate and distinctive levels. The level of insight needed to deliver the intervention with an effective plan and model is not solely addressed by knowing the sort of intervention necessary at the strategic level (deter, redirect, etc.). This is true for both reach-related concerns and confirming that the 'correct' audience(s) have been targeted and reached. There is a direct correlation between influential audience targeting and insight, and the stronger the level of understanding, the more accurate the audience targeting. This supports the idea that strategy and design phases are crucial.

Creating an efficient plan is the easiest way to address queries and concerns about impact and reach. Does the strategy's desired outcome match the goal(s) specified by the customer, the problem-set, or the stakeholder? The logical framework, proposition, and rationale created to communicate what must be done, why it must be done, and how it will be completed are referred to in this context as the strategy. In preparing for P/CVE online intervention, it is vital to consider the impact (both intended and unforeseen) of the campaign, message, or information in the public domain. What must be done and how it will be done depend primarily on strategy and how well it works. This creates the need for an adequate adversary analysis, a robust proposition/theory of change, an exceptional grasp of the creative treatment required for the CN product, and a raft of potential effects alongside the tactics/levers that will create them [11].

An effects framework can help reduce reliance on generic reach data. This framework is a set of detailed indicators that measure the impact of a campaign's specific methods, content, or messaging. It aligns with the campaign's overall goals, which were defined during the planning phase. For example, if a project aims to change behaviour, a goal might be to weaken an opponent's narrative. The effects framework then measures the target audience's behaviour changes. To calculate this change effectively, careful planning is needed. The framework should include measurable indicators that track the expected changes. More than reach metrics alone are required to assess the impact of online interventions. Therefore, more comprehensive evaluations are needed. A communication strategy is crucial in this context. With a proper assessment framework, a well-designed strategy can effectively measure its intended impact [17].

A practical assessment framework requires two essential pieces of information at the beginning:

- Desired End-State: A clear definition of the campaign's goals.
- Baseline Audience Understanding: This refers to understanding the target audience's initial level of support, sentiment, and other relevant factors regarding the addressed issue.
- This information helps develop better outcome indicators, the metrics used to measure the campaign's success. A baseline evaluation, which should be part of the assessment framework, is the most effective way to assess the required audience movement and how to track it. The baseline evaluation also allows for measuring the achievement of the desired end-state [10].

Campaigns without a well-defined assessment framework risk relying on vanity metrics, which provide little meaningful information. A clear end-state objective is crucial for establishing measurable goals, outlining strategies and tactics, and demonstrating the campaign's effectiveness. It can be ideal for an online P/CVE intervention to work in tandem with offline initiatives. At the local level, the idea of the 'audience' hides the features or meanings behind terms like 'network', 'community', and 'activism'. The issue of some audiences being 'hard to reach' is compounded by the fact that online audiences are elusive and challenging to 'read' in isolation from offline realities. This problem can be resolved by trying to base a sizable amount of a project's material on offline interactions with people who are not yet audiences. Alternatively, depending on the goals and tactics of the campaign, offline participation may or may not be necessary. It may only be valid for some learning from the offline world to be analysed and possibly factored into the strategy development phase. This context also ties into the requirement that P/CVE initiatives in the internet environment should attempt to go beyond the conventional focus group method and start including audiences during the conception, distribution, and assessment phases as active stakeholders in the whole process.

Despite their limited funding, CSOs and NGOs have more direct access to the networks and community actors that close the knowledge gap between the project and the experiences of localised diasporas. Campaigns based on openness, transparency, and trust are more likely to succeed than unverifiable ones, which may cause scepticism, pushback, and anger. The initiative has a higher chance of success over the long term the closer the offline involvement is to the online iteration. The tracking of sentiment and sentiment change follows the same rules. It cannot be stressed how important it is to comprehend the 'audience' in the online sphere while developing sensitive communications [14].

In the ideal scenario, practitioners' offline access to and engagement with audiences/communities dramatically benefits a project or campaign's measurement and impact requirements. Delivering promising interventions from a community/CSO/NGO viewpoint substantially benefits this setting because of physical access to various potential target groups. These organisations are in a position to work more directly with networks, interest groups, and people to advance project goals because of their local roots and location. A process that can add diversity to various potential target audience needs with specialised roundtables and workshop-style events that enable greater nuance and depth to be added to project design elements is made possible by community engagement rooted in offline relationships. This allows design and planning to go beyond the norm of 'focus group' testing with relatively unknown audiences. The commercial sector offers several useful avenues for CSOs to explore to strengthen their current offline engagement, even though the CSO and private sector do not always naturally fit together in certain contexts where subtleties may require more than 'cultural expertise' from a distance and instead be grounded in the real-world experience of vulnerable communities [16].

The degree to which a project's testing or pilot phase pinpoints its advantages, disadvantages, challenges, and subtleties closely correlates with the likelihood of success. Focus group conversations on a fundamental concept or idea usually comprise the first testing stage. Through these discussions, the project team and potential audience members can discover good, riskier, or more unfavourable aspects of the design. This can be improved, for example, by making straightforward adjustments to the A/B testing approach, including producing many variations of the same core material and distributing them to similar target audiences to test and evaluate their reactions to the slight variations in content and presentation. To target different potential consumers with the same core content through hashtag/metadata adjustments, the AX project (Chap. 4), which was a CSO/YouTube cooperation, used A/B testing by developing various thumbnails with different levels of intrigue.

Results of A/B testing and information on which content gets more engagement are not as easily transferable to impact measurement. Iterative approaches open to comments and recommendations from the sample audience regarding potential language and tone, as well as cultural and related nuances, provide an alternate strategy for eliciting more nuanced testing phase input. There is a claim that testing with more controversial or harsh information can make pre-existing problems worse. Despite this possible risk, this strategy also offers a way to maintain a degree of tolerance toward content modifications necessary to increase or sustain user engagement and confidence. An approach like this allows for a phenomenon called a 'network effect' to take place; audiences who are familiar with the content, having participated in testing, can then increase the reach of the final product to their online and offline networks.

The use and utilisation of technological tools play a crucial role in creating and disseminating online campaigns aimed at P/CVE. This chapter delves into examining the functions performed by social media platforms, data analytics, and machine learning algorithms in detecting and mitigating extremist content. Furthermore, this study examines the utilisation of artificial intelligence in identifying potential indications of radicalisation and evaluating the effectiveness of interventions. The chapter also discusses the possible issues associated with privacy and ethical considerations in using technological technologies. Interventions based on 'prebunking' methods are becoming more popular regarding approach and content design. Prebunking efforts also referred to as work with a broad inoculation theory approach as its central tenet, have made important design and capability advancements, moving firmly into the preventative space with interventions that see 'prevention, not cure' as the best way to engage in the P/CVE and disinformation/misinformation space. A critical development in this type of approach has been the transition from 'debunking' to 'mental armour' through prebunking content, which aims to prepare online users to recognise, assess, and make knowledgeable decisions when extremist or disinformation content appears on their screen or social media platforms.

However, there are difficulties with this approach, particularly in managing the addition of large scale and size while maintaining nuance and stratification. Future revisions of this methodology may significantly improve this component. Kurt Braddock's work offers an excellent blueprint for prebunking and inoculation activities to arise in the P/CVE space. By providing both test phase results and additional contextual criteria for the future development of this work, Google has made an effort to advance the practical application of prebunking efforts in collaboration with Braddock and others. Because their dissemination rates are insufficient to draw users' attention away from the original harmful extremist information, prebunking-based efforts may find it challenging to match the amount and tendency of extremist content.

Due to the likelihood of specific residual issues, ideas, or narrative components persisting among vulnerable or extremist online users, prebunking is thought to be a more effective technique for developing mental, emotional, and digital resilience in preventative settings. Users will be more able to identify and reject extremist-based information if enough instances are shown through prebunking/inoculation-based approaches [8, 9]. Inoculation as a concept operates in the preventative space yet can be vulnerable to being misused, overlooked, or seen as divisive by its intended target audience. For those well entrenched in the extremist mindset, issues of motivation, intention, and control play a significant role in how countermeasures work or do not work. Regardless of the tactical approach of an online P/CVE intervention/CN, a solid theoretical framework, model, and methodology are required. Later in this chapter, the RAM will be fully introduced as one of the default or go-to models for behaviour change-based work, including within P/CVE contexts.

Socialisation theory is a critical theoretical framework used in this research study. It is presented as an alternative to the P/CVE industry's reliance on the concept of radicalisation. Socialisation theory can help us understand, deliver, and measure counter-extremism efforts at strategic and operational levels. Online campaigns that use strategic communication tools and methods to achieve behaviour change need to consider the social context of the online environment. This is important because the online environment has its own social context that may differ from the target audience's offline social context.

The AX project (Chap. 4) used a rigorous testing process to develop its messaging and creative materials. This included testing branding, content style, duration, narrative, audience targeting, and sentiment analysis. The project also included an offline component where the project team brought the ideas and materials to schools in key locations in the UK. This allowed the target audience to provide feedback on the design and identify elements that were most engaging. A/B testing was used to compare different versions of the content, including more challenging versions, to see which ones were more likely to resonate with the target audience. This testing helped to inform the overall creative strategy and ensure the content was delivered in a way that would be most effective.

This can be valuable in the long run of a campaign in terms of working towards the correct success criteria, aims, and objectives. In this context, it can be asserted that further innovation with A/B testing, applying greater cultural nuance, using participant feedback for design improvement and improving measurement systems form part of what has been labelled throughout this study as an 'iterative approach'. Whilst several of these elements are already part of the design process for many P/CVE counter-narrative efforts, the quick-win nature of much of this sector's work nullifies this approach. There is, as such, room for improving the understanding of what a holistic and cutting-edge iterative approach can look like.

Adaptive delivery, which differs from reactive delivery in content tone and form, is a fundamental way offline activities can play a significant role in online intervention approaches. An element of scope for re-rendering and re-arranging narrative tone and form to better meet target audiences' needs through engagement analysis is retained in the adaptive framework for project content design. The tolerance for modifications and slight changes is undoubtedly greater where information is based on social media posts or audio programming (such as radio or podcasts). The major takeaway from the offline domain for more creative delivery formats like video, animation, serials, etc., is to ensure that the early testing phase can gather feedback on concept, innovative treatment, storyline, narrative, and tone [6]. The individual user experience still present in a group or community delivery setting is a crucial component of the offline experience. This is reflected in content that may accommodate intellectual, emotional, and personal 'touch points' from a single overarching delivery thread online. The CC project (Chap. 5) and the SR intervention strategy (Chap. 5) are two instances where this method is mentioned in the project criteria [5, 12].

Engagement, Measurement and Evaluation: Behaviour Change and Sentiment Analysis

In the relationship between online (or offline) strategic communications and P/CVE, it is typical for measurement and evaluation systems to be an essential part of the overall design and planning process when delivering a CN. Using the term strategic implies some form of intent and highlights a need to capture or measure this intent. The need to adopt such an approach stems from the central premise of online P/CVE CNs: seeking some form of behaviour change amongst 'targeted audiences'. Within the commercial sector (advertising, corporate communications), which has generically been co-opted by online P/CVE efforts as a form of benchmark, this often takes the shape of 'vanity' measures. The main goal of such indicators is to give both the delivery agent (advertising agency, government agency, CSO/NGO) and the 'client' or primary stakeholder a set of measurements (likes, shares, impressions) and the quantity of times a piece of content is 'seen' or flashed onto a screen (impressions) and then 'engaged' with, also known as a Click Through Rate (CTR), that provide a top-line perspective of the 'effectiveness' of the campaign.

These metrics also gauged the success and efficiency of the campaign or project as a whole and discussions about value for money. This practice needs more methodological development in terms of how such measurements can be viewed as part of more in-depth qualitative and quantitative aspects of behaviour change-based objectives, including a greater emphasis on capturing sentiment, longitudinal impact, and the role of other interlocutors (like the individual's offline/online ecosystem). This is especially relevant for P/CVE efforts in regards to how exactly and effectively such metrics can provide pertinent data and outcomes for highly subjective and open-ended concerns like radicalisation, deradicalisation, disengagement, and desistence. The idea of online P/CVE interventions is founded on particular presumptions about the function and potential attitudinal and behavioural impact of such interventions on audiences displaying variable degrees of receptivity to messaging, unequal access to information, emotional intelligence, and motivational facets.

While each seems like a logical metric for P/CVE campaigns, focusing solely on it can be misleading. Reach tells you how many people saw the message but not if it influenced them. P/CVE efforts should prioritise impact, not just reach. Mass audience targeting can be risky in the P/CVE field due to the controversial nature of radicalisation. When relying solely on vanity metrics like reach, specific circumstances and challenges can skew project outcomes. This highlights the importance of a well-designed assessment framework that goes beyond reach. Such a framework would consider the project's goals and target audience to develop meaningful outcome indicators.

Effective measurement of counter-radicalization (P/CVE) communication initiatives should move beyond simply measuring reach. A focus on engagement, rather than just how many people are reached, can provide more meaningful data. The type of engagement and how it supports deradicalisation efforts matter most. Reach data may not be sufficient because it needs more nuance and specificity. To assess the effectiveness of P/CVE initiatives, it is crucial to determine the most relevant evaluation standards. For example, evaluations of the UK government's 'Prevent' strategy have been criticised for lacking focus and relying on a loosely knit and often strategically disorganised approach to achieve quick wins. Effective evaluations should go beyond justifying a project's existence and avoid relying on arbitrary concepts like values or specific definitions of moderate versus extreme views [3].

The UK government's 'Contest' strategy (CT strategy) exemplifies a seemingly flawed approach to counter-extremism. Using broad and unsubstantiated evidence, it relies on a binary equivalence between social cohesion and counter-terrorism. This approach creates practical problems. For example, defining target audiences and program goals becomes difficult when social cohesion and counter-extremism efforts are combined without a nuanced understanding of the underlying issues.

More promising approaches differentiate between countering violent extremism and building resilience. The Global Forum on Counter-Terrorism's (GIFCT) framework, for instance, distinguishes between countering extremist ideology (upstream) and building resilience (downstream) to prevent radicalisation in the first place. Evaluation is also crucial for assessing the effectiveness of deradicalisation and disengagement programs. The type of program (resilience building, support services, etc.) should inform the evaluation approach used to measure success.

Effective counter-violent extremism (P/CVE) communication initiatives (CNs) require careful planning and a well-defined strategy. This strategy should consider several factors, including:

- Campaign Goals: what does the campaign aim to achieve? Is it to raise awareness, change attitudes, or encourage specific behaviours?
- Target Audience: who is the campaign trying to reach? Understanding the audience's needs and interests is crucial for developing effective messaging.
- Theory of Change: how will the campaign achieve its goals? This involves outlining the steps that will lead to the desired outcome.
- Effects Framework: how will the campaign measure its success? This framework should include specific indicators that track progress towards the campaign's goals.

Equally relevant is the issue of how such campaigns attempt to add further value to vanity metrics and, as a result, how specific interpretations of this type of data can often obscure the actual need for rich contextual data on audience engagement, reaction and sentiment. 'Calls to action' have traditionally been considered helpful indications for assessing the degree to which audiences engaged with a message, understood its premise, and then actively chose and made decisions to engage in project-specific 'desirable' actions. Relying entirely on online reach data can lead to false and biased ideas of audiences and their requirements due to advancements in social media platforms, tech tools, and a higher degree of audience lifestyle segmentation. A straightforward illustration would be to interpret someone's response to a piece of content (like, dislike, share, retweet, etc.) as a reflection of their belief system, tastes, or position on a specific subject matter. This also holds for contextual variations, as the same person may experience completely distinct emotional meanings from the same piece of knowledge provided in a particular way in different circumstances. Such concerns are essential to ensure the distinctions between extremism and violent extremism and that reach and influence matters are correctly distinguished. The issue of audience interaction with CN content is a crucial concern for P/CVE practitioners, as it sets out elements of the basis by which these campaigns and CNs position themselves as examples of effective practice and, by extension, as examples of 'good' P/CVE.

The relationship between content dissemination and audience interaction continues to be the benchmark of online P/CVE CN interventions, supported by any assumed values alignment and behaviour change. CN intervention design and narrative should be guided by the idea that genuine and frequently challenging engagement occurs beyond basic reach measures. The concept of active strategic communications essentially elicits a response beyond clicking 'like'. It moves the content and its audience into a new space of engagement that may be challenging or affirming but allows the audience to express ideas and feelings and (ideally) make positive decisions. Since deradicalisation efforts are distinctly different from resilience-building efforts in terms of tone, message, and objectives, as well as the type of action or desirable behaviour sought, it is evident that the context of the campaign heavily influences the kind of engagement that may be possible (e.g. building followers and generic metrics versus supporting a local network to improve governance structures to counter-extremist influences).

Engagement is crucial in and of itself, though, as comments and other types of communication give other users social cues regarding the material [18]. Users' comments on a video or article can change how they and other users perceive the content or message to be more accurate, so community management and engagement processes can boost campaigns' effectiveness and offer significant evaluation benefits [4]. Interventions in the online space, including P/CVE-related efforts, have started to evolve alongside a change in approach as to what type of intervention at what stage is required to have the most effective all-round impact, even if this means expectation shift from quick-wins to longer-term effects over time.

A campaign's ability to inspire critical reflection at the individual level or to use persuasive messaging to influence audiences to take specific actions that contribute to achieving larger end-state goals is greatly influenced by the narrative structure and content style used in the campaign. Both of these effects are related to the concept that content can enter and affect the end-cognitive user's space while looking for what is referred to as a 'cognitive opening' [19]. Managing unfavourable comments, pushback, or 'blowback' of content from audiences who may or may not be a part of an intended target audience is another facet of transferring offline techniques to the online arena. This situation is frequently viewed as a grey area in the P/CVE world where campaigns seek to gain the trust or credibility typically reserved for face-to-face offline interventions in deradicalisation and prevention. The grey area is often seen in a campaign's inability, lack of confidence, remit or scope to address such issues. Ironically, negative feedback presents an opportunity to address complaints, false information, or hate speech directly using tact, composure, and intellectual capability, depending on whether a campaign focuses on offering alternative narratives or some CN.

These characteristics are not typically seen in extreme narratives, usually noted for their binary nature and obstructionist tones. This does not negate that the story, structure, and tone of extremist discourses are complicated. Resistance or push-back indicates that individuals contributing to the problem set have responded to the campaign's persuasive presence. A well-crafted, acceptable reaction can effectively counter this criticism and increase your presence in the crowded online environment. The converse is true for hasty, frightened, and poorly considered answers that could unintentionally spread or indirectly support extremist messaging. This approach's viability will ultimately depend on the project's scope and context. Still, the fact remains that well-planned, tactical, direct engagement of negative feedback can give a campaign more opportunities to build trust and influence behaviour. The following section, which examines how engagement is defined through audience sentiment broadly, fits neatly into the previous section by examining audience feedback and behaviour change difficulties as both a learning point thus far in P/CVE and as a chance to develop a greater understanding of potential avenues for future consideration.

Without additional inputs at a broader scale, the function and scope of communications-based efforts to counter and address challenges like extremism remain limited. The power of effective communication rests in its capacity to apply and interpret both its presence and the presence of communication/messages coming from other actors, even though this could be construed as a loss-leader or risk. This pertains to the type, tone, quality, and volume of communication a campaign receives from its target audiences and analyses in the context of positive CN interventions.

Sentiment analysis, or identifying positive or negative sentiment in text, is the name given to this method. This endeavour identifies specific sorts of 'polarity' that can be deduced from textual analysis (e.g., positive, negative, neutral). However, sentiment analysis is a tool that interprets certain moods and emotions (such as happiness, sadness, or anger), urgency, and even goals. It does this by going beyond the straightforward process of polarity identification (e.g. interested versus not interested). These can be divided into separate processes, detailed below.

Evaluate various sentiments concerning specific desirable and undesirable criteria.

- Conduct emotion detection analysis from simple text
- Conduct aspect-based analysis of sentiment driven by particular factors or conditions,
- Perform multilingual sentiment analysis that examines variations of the same sentiment in various linguistic contexts

Real-time analysis, data sorting at scale, and using uniform criteria are key advantages of sentiment analysis as a methodology. However, there are methodological problems with how this strategy would be applied in preventative contexts, as it is challenging to determine tone from written words, particularly in large data sets and across cultural boundaries. With mid-polar phrases that depend on tone, sarcasm, culture, idioms, and context, polarity is another challenge. When there are sharp polarity shifts, sentiment analysis is a valuable tool for assessing individual or community variances, although it is frequently scale-dependent.

Using lexicon-based sentiment analysis in this context may be favoured to gain a preliminary understanding of public opinion towards CVE messaging. This approach has limitations, however, and may not capture the full range of sentiment. For example, sentiment analysis tools can have difficulty distinguishing between factual statements and opinions, and may misinterpret sarcasm or irony.

Below are some of the limitations of lexicon-based sentiment analysis:

- Difficulty distinguishing facts from opinions
- Misinterpretation of sarcasm and irony
- Inability to capture the full range of sentiment

More sophisticated approaches, such as ensuring a more nuanced and culturally balanced machine learning approach, may provide a more nuanced understanding of public opinion. It is crucial to remember that sentiment and effectiveness are not the same when it comes to effort in P/CVE contexts. The notion that participants' 'feelings' are proof that an intervention is working is one of the most prevalent validity problems with intervention effectiveness measures. That is not always the case. According to discrete emotion theory, each emotion has a unique action tendency, which means that when a person feels a particular emotion, they are compelled to behave in a specific way. For instance, if people react to an intervention with wrath or disgust, they will likely discount it psychologically. In this approach, it is crucial to comprehend (a) the feelings intervention participants have and (b) the actions those feelings lead them to take. The takeaway is that not all emotional reactions are equal and that adopting this assumption can result in ineffective intervention designs.

This implies two methods of measuring discrete emotions (or sentiments). Firstly, thought-listing techniques have a lot of potential value. Participants in a thought-listing activity are merely instructed to 'free-write' their feelings about the intervention. Participants typically get roughly 10 min to do this. Regarding what participants can write, there are no guidelines or limitations. After this thought-listing exercise, two or three coders review and categorise all of the ideas written by participants. Although coding schemes can be created in various ways, they must, at the very least, have the following measures: (1) relevant/not relevant and (2) positive/negative. Positive sentimental responses to the intervention can be inferred from ideas coded as 'relevant' and 'positive'. A conventional survey is a second technique for gauging reaction to an intervention. There are several ways to create surveys however it is advised to incorporate critical sentiments within a broader emotional scale.

Generally speaking, to question participants directly about the intervention's components, measures of beliefs and attitudes in response to intervention practices must be customised to the intervention. For instance, belief scales should enquire about specific conspiratorial beliefs if the goal of the intervention is to reduce these beliefs; if the goal is to change attitudes toward a particular minority group, an attitude scale should be developed to gauge participants' positive or negative impressions of that group. The RAM approach can quantify beliefs, attitudes, and sentiments. This theory contends that ideas about behaviour, social norms associated with the behaviour, and one's capacity to engage in the behaviour all influence behaviour. In other words, exposure to specific messages can influence attitudes toward violence, which in turn influences intentions to use violence, which influences violent behaviour through control.

Analysis that considers various contact points or nodes of narrative engagement and sentiment is perhaps the most sought-after part of project measurement and evaluation. Here, more sophisticated types of sentiment analysis, including the presentation of visuals, take centre stage. Analysis may consider the overall mood, the mood over time, the mood by rating, and the mood by subject. Different effects and sub-effects can be found and correlated to determine whether a specific tale has achieved its targeted conditions or aims by combining various contact points or measuring tools. This method is also highly effective for assessing change across multiple settings and contexts. For online therapies that only rely on the power of communication to have an impact, such a procedure has quickly become the norm.

The ability of tech tools to 'opinion mine' in ways similar to how data mining has become a regular procedure for big data sets is a significant tool in sentiment analysis. The opinion mining process, while standardising the analysis component to quantify 'impact' or results, is not without its flaws, some of which may impact the general success criteria of a project/campaign or particular narratives.

A sentiment analysis approach presents a variety of difficulties and serious risks, which pose a similar threat to counter-disinformation and P/CVE initiatives. When evaluating complicated data sets from audiences who may express their opinions differently depending on their culture or religion, subjectivity and tone are challenging to establish. This may necessitate localising some concepts. Contextual reality concerns must be taken into account when identifying polarities. Irony, humour, and sarcasm are only a few factors whose application and subsequent detection/analysis still call for more work and skill. There may be disagreement on the comparison methods and their standards regarding accurate measurements of sentiment.

Further research is needed in the underdeveloped area of interpretation and comprehension related to emojis and icons/favicons, which have fast become 'go-to' tools for expressing emotions. Such a framework's definition of 'neutral', which might be perceived as a very arbitrary option of no real meaningful value, is one of the most critical issues to understand. This relates to machine learning systems using algorithms to detect and understand data versus the accuracy of human annotators and emotions. About the problems of interpretation and measurement of sentiment change, which are nevertheless crucial indications of the possibility of behaviour modification, sentiment analysis has a wide range of prospective directions to take despite these disadvantages. It is also important to note that radicals of all stripes are getting better and better at using sentiment analysis to further their narratives. A more complex explanation of these options is introduced in the following section.

Making sure that the measurement of sentiment change entirely takes into account the distinctive nature of sentiment that is culturally specific, driven by more individualistic tones, and its origins and manifestations in ideological underpinnings is one area of the reach and impact debate that is full of rich potential. The sentiment change analysis involves both flexibility and a certain level of sophistication to read language, form, and tone in a way that is as nuanced due to its originator's (individual's own) levels of understanding when these three nodes align inside the expression of (for example) one individual. This also applies to the stricter sentiment to label in terms of positive or negative, etc., because it makes several complicated assertions that could be interpreted as affirmative, rejecting, challenging, or questioning. This is especially true for complex interventions involving audiences that might be receptive to some aspects of extremist thinking (conservative or literalist thinking) while being indifferent to other elements (worldview or violence). Such intricacy puts the subjectivity debate in sharp relief, highlights the importance of indigenous studies to P/CVE efforts, and highlights the relative inequalities between distinct groups regarding influence, power, and representation.

Applying semiotics research combined with sentiment analysis can help overcome these obstacles and produce an even more complete picture. Semiotics studies signs and symbols that place their significance and values within specific societal and cultural contexts. This makes it possible to assign interpretive meanings to particular language and expression motifs (colours, phrases, terms, icons, and symbols) depending on the cultural circumstances that initially created the value through shared experiences, tales, and communities. This is best illustrated by the fact that the term 'conservative' has a variety of connotations in many cultural, theological, and political contexts, making it dangerous to assign one overarching interpretive meaning to it. Specific colours and symbols can be used to create distinct meanings under certain circumstances (such as flags and slogans), and as a result, they can be used for quite diverse things.

Semiotics can be used at the design and planning stages and the evaluation and measurement phases to better understand potential outcomes and what these may signify [2]. The possibility that 'impact'-related issues and sentiment change are valid and can guide future needs and interventions increases with the design's degree of cultural competence. All of this advances the discussion of sentiment shift and its analysis into a broader, more tenable area, enhancing the potency of narratives across the board, not only in the online sphere. Understanding sentiment and change as phasic inputs produced throughout a project's cycle is crucial to this approach. Sentiment analysis and semiotics are instruments that help capture sentiment change.

Additionally, there are many ethical, legal, and moral considerations to understand and factor in when designing the communication product and preparing for engagement, sentiment, and interaction. These factors are significant issues but are slightly outside the scope of this research study to discuss at length here and, therefore, will be discussed in relevance to concluding findings later in this study where possible. The focus of this chapter needs to shift to a discussion of the RAM to lay the foundations for the case studies and bring in relevant considerations of audience targeting needs concerning models of behaviour change, such as the approach identified as one of the core theory frameworks for this study.

Summary of Chapter Findings and Key Learnings

This chapter has examined the emergence of online P/CVE communications and how strategic communications needs both affect and create effects (are influenced by and seek influence) for the interventions made. The discussion links broader P/CVE policy and strategy issues to the burgeoning online P/CVE space to create a more robust discussion of the state of the sector and its emerging needs. In summary, this chapter has thoroughly examined strategic communications issues and online activities within P/CVE. The discussion has emphasised the importance of strategic planning and execution in campaigns, compelling storytelling to connect with audiences, effective measurement and evaluation methods, and the crucial role of technical tools in countering online violent extremism. By comprehending these facets, policymakers, practitioners, and researchers can formulate more productive and focused strategies to tackle the intricacies of fighting violent extremism in the digital era.

The analysis starts with a broad look at the online P/CVE domain to couch the more granular sections to follow with a suitable context.

The first chapter highlighted that any communication effort with a primary purpose behind its creation is strategic. This is perhaps where the issues within the P/CVE CN space start and what this current chapter has highlighted; namely, that the mere 'desire' to counter a narrative renders the desire as a CN, yet often without a framework to pin its strategic auspices, tactical levers, and theory of change on. There is an intention at the outset, some form of tactical operation or content dissemination, and loyalty to collecting vanity metrics and 'data' to present or evidence the overall 'successes' of the effort. Many experts point to the short-term 'quick-win' nature of P/CVE CNs as why such conceptual and evaluative factors do not require in-depth work, given that the CN effort may be time, money or stakeholder support-specific [7]. This starkly contrasts the field of intervention for which RAM is most associated: health-based behaviour change [1]. Health behaviour change (HBC) communication campaigns often have a relatively short or fixed lifespan in terms of their visibility and engagement but equally incorporate longitudinal measurements of behaviour change to aid the reliability of critical findings [13].

The lifespan of such communication efforts is linked as much to material factors (time, money, etc.) as to using RAM as a behaviour change theory. A more immersive and tactile iterative approach to design, delivery and measurement needs would be a helpful pathway for the sector to develop its quality and efficacy levels further. This is the exact nexus of issues discovered through the in-depth study of P/CVE CN work in this book and if the field is to enhance its overall efficacy, reliability, and impact over time. The case study analyses that preceded this chapter aimed to assess and unpick elements within each example of socialising tactics, behaviour change, and strategic communications usage. The SoRaSCo model has offered a proper mechanism by which to unpick the specific elements of each effort and place them within an overarching conceptual and practical framework. In the final chapter, the discussion moves into the realm of capturing beneficial elements of P/CVE communications this book has unearthed and also provides scope for outlining currently underdeveloped or less productive aspects of the sector's approach.

References

1. Ajzen, I. (1991). The theory of planned behavior. *Organizational Behavior and Human Decision Processes, 50*(2), 179–211. https://doi.org/10.1016/0749-5978(91)90020-T
2. Barthes, R. (1977). *Elements of Semiology*. Hill And Wang, New York, NY.
3. Baumeister, R. F., Masicampo, E. J., & DeWall, C. N. (2009). Prosocial benefits of feeling free: Disbelief in free will increases aggression and reduces helpfulness. *Personality and Social Psychology Bulletin, 35*(2), 260–268. https://doi.org/10.1177/0146167208327217
4. Cheung, M.L., Pires, G., Rosenberger, P.J. (2020). The influence of perceived social media marketing elements on consumer–brand engagement and brand knowledge. *Asia Pacific Journal of Marketing and Logistics*, [online] 32(3), pp.695–720. https://doi.org/10.1108/apjml-04-2019-0262.

5. Frennett, R. and Dow, M. (2015). *One to One Online Interventions—A Pilot CVE Methodology*. [online] ISD. Available at: https://www.isdglobal.org/isd-publications/one-to-one-online-interventions-a-pilot-cve-methodology/ [Accessed 20 Mar. 2022].
6. Glazzard, A. (2017). *Losing the Plot: Narrative, Counter-Narrative and Violent Extremism.* [online] International Centre for Counter-Terrorism—ICCT. Available at: https://www.icct.nl/publication/losing-plot-narrative-counter-narrative-and-violent-extremism.
7. Institute for Strategic Dialogue (2016). *The Impact of Counter-Narratives.* [online] ISD. Available at: https://www.isdglobal.org/isd-publications/the-impact-of-counter-narratives/#:~:text=ISD [Accessed 19 Nov. 2022].
8. Jigsaw. (2018). *Medium.* [online] Medium. Available at: https://medium.com/jigsaw/hate-clusters-spread-disinformation-across-social-media-995196515ca5. [Accessed 18 Mar. 2023].
9. Jigsaw. (2022). *Medium.* [online] Medium. Available at: https://medium.com/jigsaw/prebunking-anti-vaccine-narratives-an-effective-alternative-to-debunking-individual-false-claims-78f0047a8b47. [Accessed 21 Jul. 2022].
10. Lee, B. J. (2018). Informal countermessaging: The potential and perils of informal online countermessaging. *Studies in Conflict & Terrorism, 42*(1–2), 161–177. https://doi.org/10.1080/1057610x.2018.1513697
11. Lee, J. (2020). 5 Practical tips to step up your audience targeting game. *Search Engine Journal*, [online]. Available at: https://www.searchenginejournal.com/5-practical-tips-to-step-up-your-audience-targetinggame/373428/#close. [Accessed 18 Mar. 2023].
12. Moonshot CVE. (2021). *Canada Redirect—Final Report—Moonshot.* [online] moonshotteam.com. Available at: https://moonshotteam.com/resource/canada-redirect-final-report/ [Accessed 19 May 2022].
13. Ogilvy, D. (1995). *Ogilvy on Advertising*. Prion.
14. Riddle-Nunn, M. (2018). Tips and tools for understanding your online audience. *Mostly Serious*, [online]. Available at: https://www.mostlyserious.io/news-updates/understanding-your-online-audience. [Accessed 19 May 2022].
15. Ritzmann, Wouterse and Verdegaal. (2019). Product of the RAN centre of excellence authored. Available at: https://home-affairs.ec.europa.eu/system/files/2019-12/ran_cn_academy_creating_implementing_effective_campaigns_brussels_14-15112019_en.pdf [accessed 13 April, 2022]
16. Sen, A. (2020). *7 Offline Customer Engagement Strategies You Didn't Know About | Blog.* [online] involve.me. Available at: https://www.involve.me/blog/offline-customer-engagement-strategies [Accessed 19 Mar. 2024].
17. Tariq, H. (2021). *Council Post: Five Components of A Successful Strategic Communications Plan.* [online] Forbes. Available at: https://www.forbes.com/sites/forbescommunicationscouncil/2021/06/22/five-components-of-a-successful-strategic-communications-plan/?sh=326789e05813. [Accessed 19 Mar. 2022].
18. Walsh, L., Hyett, N., Juniper, N., Li, C., Rodier, S., & Hill, S. (2021). The use of social media as a tool for stakeholder engagement in health service design and quality improvement: A scoping review. *Digital Health, 7*, 205520762199687. https://doi.org/10.1177/2055207621996870
19. Wiktorowicz, Q. (2005). *Radical Islam rising : Muslim extremism in the West*. Rowman & Littlefield Publishers.

Chapter 7
Conclusion

Abstract This concluding chapter synthesizes the study's findings on the efficacy and strategic considerations of online counter-extremism (P/CVE) communications. The research identifies three core issues by applying a hybrid theoretical model (SoRaSCo)—which integrates socialisation theory, reasoned action behaviour-change models, and strategic communications. Firstly, over-reliance on radicalisation discourse skews P/CVE work toward psychological frameworks, overlooking crucial social and contextual factors shaping extremist engagement. Secondly, current approaches often fail to incorporate socialisation processes that influence norm formation, vulnerability, and audience receptivity to counter-narratives. Thirdly, these conceptual and methodological gaps blur the strategic intent and clarity needed to design and measure effective counter-narratives (CNs). By examining three case studies, the study demonstrates how nuanced, socially grounded approaches (exemplified by the AX project) outperform generic or metric-driven efforts (as seen in CC and SR) that prioritise reach over meaningful behaviour change. The research highlights the necessity of embracing deeper cultural, normative, and social insights to guide content design, delivery, and evaluation. Ultimately, the chapter argues that P/CVE communications must move beyond simplistic radicalisation models and short-term "quick wins" toward an integrated, theoretically rich, iterative strategic communications practice capable of achieving lasting preventive and behaviour-change outcomes.

Keywords CVE—countering violent extremism · PVE—preventing violent extremism · Counter extremism · Communication · Strategic communication

This book aimed to understand the effectiveness of strategic communications in countering violent extremism online. Socialisation theory, radicalisation and deradicalisation discourses, communications studies and behaviour change models (RAM) were reviewed and applied to the research parameters to achieve this aim. These theories and approaches were combined to create the SoRaSCo model, which enabled a detailed and granular analysis of the tools, tactics, nuances and gaps in current P/

M. Zamir and C. Kaunert, *Strategic Communications' Role in Counter-Terrorism*,
Advanced Sciences and Technologies for Security Applications,
https://doi.org/10.1007/978-3-031-89110-6_7

CVE online efforts. Below are the key concluding findings of this study and related areas of interest.

Three subsets of critical findings emerged after analysing three specific online P/CVE efforts using the 'socialisation and reasoned action strategic communications model' (SoRaSCo) designed for this research study. Firstly, P/CVE efforts are too reliant on radicalisation discourses, which are, in turn, over-reliant on psychological understandings of extremism and violence. Secondly, P/CVE online efforts overlook the role and presence of socialisation processes in engagement, the willingness to engage, and the formation of tangible outcomes for these engagements. Thirdly, the first two issues combine to cloud the role of strategic communications in such efforts. As a result, clear objectives for the intervention impact of persuasion-based inputs and the use of appropriate CNs are under-developed. By analytically discussing these matters for the aid of this book, these elements will be presented in this chapter as a way of concluding arguments and future directions.

Original Contribution and Significance

The critical contribution of this study to the aims of the book and the extant literature within terrorism studies and P/CVE workflows is the lack of attention paid to the extrinsic and intrinsic links between socialisation theory, behaviour change models, and the role of strategic communications. The findings suggest too much weight has been apportioned to the cognitive element of pathways into extremism at the expense of a more thorough empirical understanding of social dynamics, social agency and social action. Moreover, in terms of intrinsic links, this study has found that communication of a fundamental or strategic nature relies heavily on the presence of existing socialising agents within an individual's mindset to both couch and develop normative perceptions of new information based on intention, motivation, and perceptions of behavioural control. In other words, behaviour is predicated primarily around socialising factors specific to the individual's context. If these factors are understood in-depth before designing and disseminating P/CVE CNs, they can be utilised alongside cognitive factors outlined in the RAM. This offers any CN an opportunity to develop the narrative elements of an intervention with a social and cognitive-behavioural focus.

In doing so, the communicative element of such an effort can create strategic intent along with social and behavioural effects in tandem. This results in strategic communications that seek to change behaviour through socialising features manifested in behavioural change. The findings of this research study indicate a disparity between the underlying 'aims' of online P/CVE CNs and approaches taken within the offline sphere of prevention or P/CVE efforts. Whilst a detailed comparative analysis of both spheres was beyond the scope of the research aims, the critical distinction appears to rest in the virtual nature of online CNs, which are a step removed from the social reality of any intended user and the social context-heavy setting of offline interventions. The example of the UK's Prevent strategy approach via the Channel process of intervention support for radicalised individuals contains relevant factors, detailed below.

- The use of traditional sociology and psychology tools, such as personality and typology formation
- Assessment of character traits in line with background contextual factors
- Personal experience of Jihadi narrative/extremism implied charisma and credibility of the Intervention Provider

These form crucial elements of how an intervention is perceived and how risk-management matters are addressed in the UK's Prevent Strategy Channel project. The focus of such processes is still very much in the cognitive domain, with behavioural facets confirming a certain mindset or mentality being present and then altered via intervention in the social setting. The difference with the offline approach is the social setting and the dynamics this creates during an intervention. This contrast is evidenced through a more robust risk assessment and management approach to offline interventions. The online sphere of P/CVE efforts bases much of its moral and ethical grounding on the 'do no harm' position used in health management efforts, among others. However, as no consensus exists as to an exact definition of the principle, this approach takes on too many diverse interpretations within P/CVE work. As a result, campaigns and narratives veer towards 'safer' content and ideas rather than tackling the more challenging elements of extremist narratives and ideology. The 'quick-win or risk-averse approach limits the ability to achieve the strategic intent of the CN and leads to measurement tools which miss the social, behavioural, and contextual elements of what could constitute an 'effective' intervention.

The research findings of this book also reveal that such interventions and their methods and tactics need to be adaptive in practice to cope with emerging risks and evolving threats. Regardless of this point, learnings from this research also suggest that without key identifiable objectives being part of the intervention plan, the process can have potentially harmful or, for want of a better term, zero-sum (a stale mate situation) consequences for P/CVE needs. The primary aim of any intervention in this context is to reduce the likelihood of a person acting violently upon their beliefs, views, and ideas through radicalisation. This is the case for both online and offline versions. The term radicalisation creates both conceptual and practical confusion in agreeing on what such input should look like and what it should contain.

It also further confuses strategic planning and delivery considerations when using mass strategic communications tactics on large target audiences. The apparent ease with which entire conceptual models of radicalisation have been built with explicit assumptions about Muslim populations or about the role 'Islam'/faith plays in the radicalisation 'process', for example, points to matters already discussed in the literature review, but which require a far more nuanced and sensitive approach to be taken. These findings are more evident when the key results of the case studies analysed in this research are combined.

General and Specific Results of the Case Studies

Of the three case studies reviewed, the AX project bears the hallmarks of nuanced design elements, explicit use of general and specific narrative persuasion techniques, subtle symbolism, and a high-risk strategy of directly challenging the problem narrative. Animation, segmented audience targeting, provocative narrative tone, and storytelling elements have been merged effectively with a strategic intent seeking definable outcomes. The delivery of this project contained an ability to challenge extremist narratives in an aggressive and detached manner, allowing for different stages of the extremist narrative to be unpicked, from curiosity to clear intent to support extremism. In a style more akin to the offline intervention model (multi-layered narrative design and nuanced elements of social, cultural and historical references), AX was able to engage at an individual level yet offer much broader input using a distinct understanding of the social world of extremism and its pathways to reach the 'vulnerable' online user. AX delivered creativity, innovation, credible messaging, and challenge through its narrative delivery, which appears to have created conducive conditions for social, cognitive, and cultural behaviour shifts. The project's strategic communications and behaviour change modelling match the RAM approach and research framework of this actual study (SoRaSCo). Based on the AX case study, the mix of social, cognitive, and cultural elements suggests the SoRaSCo model has mileage within P/CVE efforts as a framework for future design, research and development needs.

The CC project based its conceptual premise along similar trajectories, swapping animation for direct messaging and using 'credible messengers' to engage at-risk users. Identifying the risk of radicalisation via a review of users' social media footprints points to understanding the nuance and sensitivity needed in this sphere. Utilising former extremists familiar with the intervention area suggests the project considered who was best placed to engage and deliver counter-messaging. The project appears to have become more risk-averse with the issue of using more provocative or real-time engagement techniques to build actual credibility as opposed to assumed credibility on the part of the intervention providers and their relationship with recipients.

There are no results that provide evidence of the social dynamics and processes at work in the mindset of a recipient during an intervention and how these offered any chance of behavioural change. The engagement tactic comes across as 'hit and hope', mimicking a cold-calling sales and marketing approach. Such a tactic does not possess the in-built capacity to engage background contextual factors, intention, norms, attitude, and perceived behavioural control of the RAM approach to behaviour change. These factors thus become anomalous as opposed to facets which can be explored and affected throughout engagement and trust building.

The SR approach relies on two key elements: targeting the right users and eliciting a response from a targeted offer of alternative content, creating the redirect premise. Using search algorithms to identify appropriate users is a practical initial step. Still, it must be suggested that this approach can only consider critical social and personal dynamics and factors at a basic rudimentary level. Whilst it is accepted that the online domain is not conducive to seeking out such intricacies, there are still ways and methods to build comprehensive digital footprints of these same dynamics during an information-gathering and data-analysing phase. Not doing so again points to the same inherent weakness discovered with CC. It is also difficult to overlook a possible 'double negative' of a user engaging with the redirected content to entrench extremist thinking further, as it allows for confirmation bias to creep in very early into the redirect process. In essence, nothing stops the redirected content from being exploited by extremists. This raises questions regarding the project's strategic intent and what actual outcomes have been planned for. As evidenced in this study, strategic communications rests on the ability to design communications with clear and measurable goals in mind.

A weakness of SR appears to be an over-reliance on vanity metrics, which do not translate well into the SoRaSCo model of this study. Such an approach severely limits RAM's behaviour change auspices and almost totally ignores social, background, and contextual factors. The project leans more towards a tech-based quick-win approach, which could appease specific stakeholder sectors within P/CVE that are trying to quell and address extremist content online. Several developmental and research areas could enhance the intent to include a more longitudinal and nuanced understanding of the user position when engaging with and supporting extremist content. These case study findings offer broader learnings that may be useful for future research efforts.

Within the CN arena, two clear distinctions emerge between content that seeks to directly challenge or undermine extremist narratives and content that aims to provide some alternative narrative, redirection, prevention, or inoculation. There can be situations where certain CN efforts cross these terms and attempt combined elements of communication. Although such distinctions appear relatively prominent, the auspices of each strand remain closely tied to a fundamentally communications-led approach to counter-extremism. Communication and its strategic variant sit at the heart of online and offline efforts within this industry. CNs of any variety rely on the power and potential of communication to contribute towards a goal. The link between this factor and strategy is where strategic communications considerations enter the picture. The type of CN needed is, in this guise, heavily dependent on the plan and end-state objective the product is seeking to achieve.

Equally, this feature (strategic intent and framework) also gives observers a chance to analyse the CN's product, form, and content, given essential elements of what the content seeks to convey should technically be discernible to the trained communications expert in the P/CVE context. Theory framework links to the RAM and socialisation have allowed this study to delve into the assumptions used to create 'strategic CNs' and what the design and outcomes of such efforts indicate are the possible benefits or risks for P/CVE of this type of approach or methodology. In essence, it has become apparent during this study that when looking at online P/CVE efforts, these can be labelled as strategic CNs for apparent reasons.

This term merges the intent, end-state objective needs, design elements, narrative components, and counter-position of such work about the proliferation of extremist narratives in the same domain with varying degrees of success. What remains an area of sparse research, however, and something this study has also encountered, is the issue of a robust evaluative framework to be developed for consistently measuring the effects and outcomes of such communications. In arriving at this point, issues around the limitations of this specific study do fall into the framework and measurement domains. It is prudent to discuss these in this concluding chapter.

Limitations

This book sought to understand and assess the role and effectiveness of online P/CVE CN efforts. Much of this work uses strategic communications campaigns to engage and persuasively seek to alter the behaviour of people identified as vulnerable to extremist narratives. During the research and information-gathering phase, some fundamental limitations emerged that have affected the study's ability to pursue the stated aims fully. These will be discussed by breaking down specific areas that have come to life as limitations, including issues with the extant literature, theory framework, methodology and findings.

Extant Literature: An explicit limitation of the study was the need for more available academic research in the field of P/CVE strategic communications before this project. Whilst terrorism studies literature is vast and expansive in conceptual depth, the field of counter-extremism communication is the exact opposite. As a result, much of the literature that is even remotely relevant to this topic stems from radicalisation discourse, which has become over-reliant on a psychological rendering of the phenomena it seeks to understand. In attempting to assess the role of communication in preventing extremism then, it was hoped that literature from sociological perspectives would also be a contributor to the research aims, given that communication is a critical social process. However, this was not the case, and the study required a more innovative research approach.

Theory Framework: Given the lack of relevant source material on in-depth evaluations of P/CVE counter-narrative efforts, the research phase instead focused on seeking to develop a basic theoretical framework that could feasibly address the critical facets of P/CVE strategic communications in the online domain, message, messenger, and recipient. Three vital conceptual parameters were identified when broadening these elements: narrative, engagement and behaviour change. This allowed for a hybrid or combined theory framework to be applied to this book. The base-conceptual level of all three elements was identified as being linked to socialisation-oriented processes.

This is because the narrative, engagement with narrative, and influence or persuasion as a result of this (behaviour change) are all elements of social dynamics occurring in normative and values-based exchanges, i.e., socialisation. Any communicative effort by nature is based on some form of the social, transactional process via a socialising agent, context, and intended shift in behaviour, acceptance or acknowledgement. Given that P/CVE strategic communications are associated with CT policy and strategy contexts, the socialisation approach required further enhancement to adequately address distinct components of the sector's efforts.

Methodology: The required enhancements are related to a need to understand precisely what constituted strategic communications in this field and to be able to assess the extent to which behaviour change could be measured. The theory framework was built using socialisation and political socialisation concepts as foundation layers. Communication studies elements are superimposed layers, added to a suitable behaviour change model (RAM), linking communication to influence and persuasion expectations. They created a 'Socialisation and Reasoned Action Strategic Communications' model (SoRaSCo). This model required a practical methodological approach to glean the most detailed results from analysing P/CVE strategic communications efforts. A case study-based analysis of P/CVE CN efforts was chosen, which could incorporate the theory framework and narrative analysis and assess impact/outcome issues. However, a fundamental limitation of this approach was a lack of available data about P/CVE CNs, limiting the number of case studies being studied, and no comparative or in-depth quantitative study to build out learning in different ways. The COVID-19 pandemic and access issues compounded some of these factors, and the decision was taken to limit the research parameters to three case studies and to analyse counter-Jihadi efforts at the expense of being able to look at Far-Right CN efforts. The researcher's previous experience meant that counter-Jihadi efforts and academic material related to the subject were more accessible.

Findings: With three case studies assessed, however, more limitations emerged with publicly available information. Only one case study (AX) enabled a thorough and conceptually rich analysis. This case study was analysed using socialisation, strategic communications tools and the RAM approach to discuss its pros and cons in depth. CC and SR were analysed using a slightly trimmed model version. This is something that affected the overall ability of the study to offer more comprehensive findings.

Challenges to Existing Perceptions from the Research Findings

Whilst the research project's limitations have impacted the overall depth of the findings, several interlinked issues have been unearthed through the case studies and broader assessment of the P/CVE CN space. In essence, this research sought to understand how well the frameworks of socialisation, communications studies, and the RAM approach apply to online CN work, with all the variables and contested and conflated issues already mentioned. The central hypothesis of this research effort is that, whilst CNs are needed to combat the narrative component of extremist ideologies, the majority of such efforts lack conceptual clarity and adequate strategic intent for differing reasons. To explain why this proposition has been made, it is essential to evaluate the type of strategic communications CNs represent against more obvious counter-extremism/terrorism strategic communications. This will enable the main findings to ascertain what is strategic about CNs and how this can be evidenced appropriately to push the knowledge gaps and practical considerations for CNs further as a distinct discipline of the strategic communications field.

At the baseline of this argument is the claim that radicalisation remains a highly contested concept and lacks the necessary clarity to create and deliver CNs [12]. Added to this is the issue that, whilst many CNs are designed for strategic reasons, the campaigns are often based on quick wins and short-termism. It is usually hard to explain the strategic intent behind a campaign, the behavioural change proposition(s), and how the outcomes provide tangible evidence of behaviour change or 'prevention' success. Being able to identify the need for a CN does not necessarily make it easy to determine how a particular CN can counter an extremist narrative and who or what mechanism is most suited to deliver this CN. The prevailing approach does, however, provide scope to seek to establish how a piece of engagement, content, or communication can be classified as a CN within the online P/CVE space. In this regard, such efforts contain specificity elements compared to non-P/CVE efforts. It may also point towards a better understanding of what CNs are trying to counter. However, the term 'counter-narrative', like 'radicalisation' and 'extremism', suffers from being too broad or seeking to be overly specific in its definitional context and its applied environment.

These terms almost appear to be 'catch-all' lexical tools aimed at satiating policy and political needs. What this does is band together relatively disparate, mysterious, and often stand-alone communication-based efforts by various State and 'civil society' actors to populate a corpus of content that, upon detailed examination, lacks specificity in regard to any noticeable robust framework or conceptual alignment [11]. This anomalous feature of the term is brought into more stark contrast when the previous discussions on measurement and evaluation needs are considered. A brief look at the origins of the term 'counter-narrative' may help offer the baseline analysis needed to delve into what the practice suggests it is seeking to be and do within the online P/CVE communications space. This is anticipated to assist the process of outlining the relational dynamics between strategic communications and P/CVE CNs, laying the groundwork for the central claim of this research exercise, namely that the majority of these efforts are counter-intuitive due to a lack of a robust conceptual framework and weak understanding of the components needed to challenge extremist narratives (in this case Jihadi) effectively.

Even a cursory glance at the term 'counter-narrative' would reveal its contextual link to another narrative that is seen to need to counter. In essence, this link shows the presence of a 'master-narrative' or dominant narrative, to which and often under which, emerges some form of intended (strategic?) counter-position in the form of narrative challenge [2, 13]. CNs, therefore, exist due to and as a result of more dominant master narratives being able to garner, foster, and imbue cognitive, behavioural, and social alignment among populations. As Andrews and Bamberg point out, these processes occur over time and through the mass internalisation of this master narrative [2]. A CN is conceptually always seeking to counter another narrative. Sadly, this apparent position becomes conflated due to the various applications of the term CN in the social, policy, and organisational worlds, not to mention the world of social movements and activism generally. This broadening of the uses of the term then creates definitional, moral, and social dilemmas as to whom, for what and why such narratives are created and with what ultimate end-state objective. A routine examination of the science of signs and symbols through semiotics reveals that objects, words and phrases carry literal and implied meanings through cultural and linguistic frames of meaning. In other words, language carries different meanings, some inherent or literal and some created through culture.

A CN, therefore, must have a meaning that is understood within a specific cultural and social framework for it to be a counter-position. Otherwise, it would not require the term 'counter' to be placed before the term 'narrative'. This means both master narratives and CNs are competing for actual or perceived dominance as stories. As such, it can be argued that stories constantly evolve, being reordered, re-rendered and repurposed to suit specific strategic, social, and personal needs. This is also highly relevant to socialisation, which codifies normative and value statements or propositions through education, familial structures, workplaces, social networks, culture and political structures [4, 9]. Narratives, therefore, are the conduit and battleground for how norms and values are transmitted in any given social system. Furthermore, the 'post-truth' era, which seems to have conceptually preceded the post-modern era, offers a stunning portrayal of a constant juxtaposition between established structures of authority (nation-states) as the gatekeepers of desirable norms and values and the emergence of social movements that reject both the power base of the nation-state and the narratives used in the protection and promotion of defence of such structures [4, 14].

As the issue of globalisation and related factors were discussed in the study, the rise of fragmented power structures, strong-man leadership, autocratic rule from within established democratic systems, and proliferation of a durable form of global disorder means that the presence of any narrative is nowadays open to challenge and rejection, regardless of what the purveyors of this narrative believe to be factual, moral, and 'right'. The mere presence of the culture of 'alternative facts' means that anything can be challenged, and perhaps the ability to 'challenge' is where the P/CVE space sits morally, yet how then does the notion of a P/CVE CN fit into the lexical, social, and political reality just described above? How can P/CVE CNs claim to be countering narratives that themselves counter to existing dominant structures? Does this mean they are countering the counter? Some of these issues were beyond the scope of this book but are pertinent to the discourses around security, prevention, and authority in the digital age.

Research Limitations and How These Findings Can Be Built Upon

The need for a more significant application of innovation with qualitative and quantitative methods, comparative A-B testing for all counternarrative efforts, and more iterative approaches to design, piloting and audience segmentation through better sample group usage are vital areas that require further detailed research. Related to these avenues of research, there is a need for additional work around the lexical issues associated with radicalisation and how socialisation theory could feasibly better explain the entire journey of extremism, CN engagement, and post-extremism contexts. There are gaps in how the behaviour change needs of P/CVE CNs are understood and designed within the content of communications campaigns. Further research is needed into the link between narrative, normalisation, and impact or behavioural control indicators.

These issues appear predicated on developing alternative approaches to the psychology-heavy emphasis of much terrorism studies discourse. This may require a paradigm shift to some extent concerning how the role of socialisation affects both communicative and interpretive aspects of P/CVE delivery. A vital element of this approach could be to enhance the use and depth of qualitative methods regarding understanding audience sentiment and reaction and extracting more iterative approaches to design and delivery needs. This study has demonstrated the ability of the RAM approach to include socialisation-based processes, iterative development needs and more nuance to cultural stimuli understanding as elements capable of making strategic communication more effective within P/CVE contexts. These additive elements combine within the SoRaSCo model and offer scope for further development within this field of study.

The research findings of this project have revealed the generic 'prevention' model, commonly known in the European context as 'prevent' and 'intervention' in the US context (although far less entrenched in CT/extremism work in the US), have been applied to counter-extremism issues via an intrinsic link to mental health, drug addiction and harm reduction models of health management and crime prevention fields. Whilst some commonality of epistemology may appear on the surface between such fields and counter-extremism, deeper analysis reveals that both the concept of radicalisation and this prevention model have created unrealistic success criteria and unmeasurable metrics concerning the nature of extremist thinking, affiliation, and action, all of which seem to be from the subjective sphere of human action and agency. Capturing this subjectivity by applying learnings and assumptions from within the health management sphere and gang prevention has been a difficult challenge for P/CVE.

This research has found that, whilst radicalisation is a fertile discourse in specific differentiation or lines of demarcation contexts, the concept of socialisation and political socialisation, in particular, address the root causes of why people seek out extremist narratives and ideologies. The highly nuanced and subjective nature of extremism requires an adaptive, multi-layered, and multi-disciplined approach to be amalgamated into a niche prevention of extremism discourse. This is especially relevant to CNs and how strategic needs should be addressed through communication and engagement with such content. This discussion has been discussed in depth in the preceding sections. Still, for now, it is important to outline why these outlier models have not been effective in preventing extremist recruitment and narratives from spreading.

Two main issues affect the approach in question. Mental illness/addiction assumption models do not take the ideological component into account. Non-violent extremism is not something that can be tackled through cognitive interventions alone and is equally an issue not suited to behaviour change models either, without clear and concise cognisance of the social dynamics within cultures that drive much of the inherent meanings and attachments so many CNs seek to engage. Without the clear presence of socialisation processes being part of any behaviour change effort, these attempts run the risk of anecdotal and vanity metric-based outcomes that do not tackle the drivers of extremism or the reluctance to show motivation for change among culturally resistant discourses. The approach that ties extremism to issues related to mental health cannot extrapolate itself from notions of extremists being people who have some form of mental or cognitive issue/deficiency, which has facilitated the onset of extremism. Within the harm reduction and drug misuse prevention spheres, the act of self-harm can be addressed through intervention on a material element (drugs/medication) where relevant and a behavioural element (self-harm/harm reduction) where appropriate.

Extremism is a lexical term, an ideological position, a relative sentiment, and a social issue with a relational link to the terms moderate or mainstream as a value proposition. It can range from an opinion to a socio-political narrative, from a world-view to a sentiment expressed in the heat of the moment. The idea is that 'intervening' through an approach based on harm reduction essentially reduces extremism to a cognitive condition that can be 'treated' through psychological means. Added to this assertion is the issue of assuming at the outset that extremism can also be altered through behaviour change when extremism may never technically manifest as behaviour or action in many cases. How, then, can a behaviour that is not manifest and is instead understood by the incumbent as a value proposition be 'changed'? This is perhaps where the inoculation method championed by Braddock has gained traction in recent times, as it posits an understanding of building mental armour that can protect an individual from harmful narratives and also enable them to identify such themes much earlier and without detriment [7].

This mental health/illness model approach discussed in the first chapter also does not possess the literary, narratology-based and emotional attachment elements of being able to tackle ideological extremism. To challenge the ideological aspect of extremism, putting forward alternatives that do not address the motivational, aspirational, and social bond-building offering suggests the approach has not sufficiently understood what extremism brings to the individual employing creating purpose. This is where the RAM and theory of socialisation are essential as theoretical frameworks within this research effort. The rationale is based against the backdrop of the mental illness/medical and harm reduction models being the drivers behind the rise of P/CVE CN, which based much of their understanding on some of the assumptions highlighted in this sub-section.

Tying these components together (harm reduction and behaviour change) offers the required segue into assessing the assumptions behind the perceived use of strategic communications within P/CVE. Technically, this could help answer the question: what is P/CVE CNs trying to achieve, and how effectively are strategic communications principles applied to this work? Based on these initial explorations, they are using the medium of strategic communications to attempt cognitive and behavioural modification via creating CNs of people deemed at risk of being radicalised or those who are already seen to be extreme in worldview and through social network association. The use of the term CN seemingly has some inherent flaws in its application within this sphere of work, given it is acceptable to ask how a CN challenges an existing CN that was intended to challenge a dominant master narrative. Explanations for why this odd scenario has emerged are needed, and the following discussion element will seek to expand on this point of debate.

It seems there is no universal consensus in a formal sense as to an agreed understanding of what a CN in the P/CVE setting means. This is partly due to the unfortunate positioning of P/CVE within state-led security issues, which means that this work sector will always be subject to theoretical, ideological, activist, and methodological critique. The very nature and purpose of the nation-state model of Western liberal democracies have come under more and more scrutiny, with the devolution of robust power narratives and erosion in terms of legitimacy and authority in the last two decades. Inherent within this scrutiny, and something unable to escape attention, is that post-modernism as both a theory and social reality has very much become a dominant force for narrative, ideation, and social action or human agency within 21st-century rubrics of meaning [5, 8].

Because of the relationship between post-modernism, the role of the nation-state, and the concept of CNs, such a set of interlinked issues emerges with P/CVE discourse or critique. It may be the case that P/CVE's position in this regard is weak, or it can also be argued that P/CVE is itself conceptually weak because it means something lexically and semiotically that does not quite add up, especially in the CN context, which is the focus of this study. CNs are confusing because of where they originate from, what they claim they are trying to counter and who they apparently 'represent'.

In a traditional social theory sense (structuralism, functionalism, etc.), the nation-state and its apparatus, status and sense of 'self' would be tied to master or 'meta' narratives, woven into ideas about authority, citizenship and freedom, rights and responsibilities, and welfare and security [16]. These meta-narratives have taken centuries to develop and become part of the fabric of civilised societies. Key to these is the idea and acceptance that the State has both authority and legitimacy regarding security and security needs. Any narrative that comes along and challenges the dominant master/meta-narrative of the nation-state concept be it through any sub-mechanism (education, healthcare, economics, freedom of expression), would be classified, identified and perceived as a CN. The argument is that such CNs respond to the meta-narrative and some form of post-enlightenment challenge to the accepted 'social order' or, using Antonio Gramsci's term, 'hegemonic order' [6].

The issue with repurposing any lexical, literary, and socially grounded term is that recipients will not make the direct connection without exceptional strategic foresight and planning, which can lead to practical, ideological, and conceptual confusion. A critical issue that has emerged from within this research focus has been that the actual narrative of P/CVE has suffered from this same lack of adequate and tangible ownership of its auspices. This means that P/CVE CNs are technically part of this same ownership, accountability, and transparency vacuum. CNs do not quite come across as CNs because P/CVE does not possess the required credibility and authority to facilitate such an acceptance. The sector should ideally seek to change this, and, somewhat ironically, socialisation theory may be a valuable tool in this endeavour.

Socialisation theory possesses an inherent capacity to be re-purposed to meet changing social realities. This theory should sit at the heart of P/CVE design and delivery, alongside an apparent acceptance that behaviour change is inextricably linked to motivational, situational, and emotive needs. It can be argued that, despite exhaustive efforts to evidence the outcome of P/CVE CN efforts, if an individual wants to leave extremism behind, they will find the motivation and pathway to attempt this, using processes and inputs unique to the situation. P/CVE efforts need to re-imagine the 'optimal moments' of prevention and change dynamics by tapping into these processes from cognitive and social vantage points.

The re-emergence of populism, otherisation, strongman leadership, erosion of trust in state structures and processes, alongside the rise of alternative 'facts', disinformation and failure to include or value the academic, social, and cultural facets of minority communities, all add to the maelstrom of potential pitfalls for a sector already found to be deficient in communicating exactly 'what' it stands for and for 'whom' it is standing [10]. At strategic, operational, and evaluation levels, P/CVE needs behaviour change, leading to a paradigm shift. Perhaps the most telling way to outline these pressing issues for P/CVE is to utilise the following statements, derived from an article titled 'Gone to Waste, the P/CVE Industry after 9/11' by Lydia Wilson.

> *One young researcher is presenting findings from a U.N.-funded project to understand why children in Iraq joined armed groups, including the Islamic State. The researcher presents a graph to dress up his work as 'science'. The horizontal axis is labelled 'extremism', and the vertical axis is called 'radicalisation'. The audience exchanges confused glances. Undeterred, the young man goes on, linking the four quadrants of his graph to differing tendencies toward violence. He provides no data or causal explanation to justify this link between violence and belief.*
>
> *Eventually, someone asks: "what exactly is the difference between extremism and radicalisation?*
>
> *Extremism had to go on the horizontal axis, from left to right, because we often talk of politics in terms of left and right", he says. Radicalism had to go on the vertical axis, bottom to top, because the word comes from the word 'radix', meaning 'root', and therefore grows from the ground up. The answer leaves the room even more mystified. It is unclear how an index as subjective and arbitrary as this could help explain the problem of children joining violent groups. "But what's the actual difference, in the real world?" someone asks.* [15]

This research project aimed to establish a conceptual foundation for P/CVE strategic communications. Whilst unable to measure several examples of such efforts, the SoRaSCo model offers some hope for how the sector can move forward in its understanding. However, overall, it highlights some limitations within the P/CVE industry more than 'instant fixes'. The industry relies on unsubstantiated theories and commercial practices rather than rigorous research. An overreliance on psychology as a catch-all solution and applying arbitrary values to complex social issues are areas for improvement. Substantial and more robust research is needed to build out core conceptual frameworks and their practical application for communications-based efforts.

Ironically, the Reasoned Actioned Model (RAM) and socialisation theory can be applied to the P/CVE sector. These frameworks help explain the predictable, repetitive nature of P/CVE approaches since 9/11. The sector resists new ideas that could improve its understanding and practices. In essence, P/CVE efforts urgently need conceptual and practical reform. A more nuanced, iterative, and social impact approach will likely be more effective.

References

1. Baker-Beall, C., Heath-Kelly, C., & Jarvis, L. (2016). *Counter-radicalisation : Critical perspectives*. Routledge.
2. Bamberg, M.G.W., & Andrews, M. (2004). *Considering counter narratives narrating, resisting, making sense*. Amsterdam [U.A.]. J. Benjamins Birmingham, Al, Usa Ebsco Industries, Inc.
3. Bauman, Z. (2000). *Liquid modernity*. Polity Press.
4. Bauman, Z. (2006). *Liquid fear*. Cambridge, UK; Malden, Ma: Polity Press.
5. Beck, U. (2009). *World at risk*. Polity.
6. Best, S. (2003). *A beginner's guide to social theory*. Sage.
7. Braddock, K. (2020). *Weaponized words: the strategic role of persuasion in violent radicalization and counter... -radicalization*. Cambridge Univ Press.
8. Fukuyama, F. (2019). *Identity: Contemporary identity politics and the struggle for recognition*. Profile Books.
9. Giddens, A. (1990). *The consequences of modernity*. Polity Press.
10. GIFCT (2021). Content-sharing algorithms, processes, and positive interventions working group part 2: Positive interventions, [online]. Available at: https://gifct.org/wp-content/uploads/2021/07/GIFCT-CAPI2-2021.pdf. Accessed 18 Dec. 2022.
11. Glazzard, A. (2017). *Losing the plot: narrative, counter-narrative and violent extremism*. [online] International Centre for Counter-Terrorism—ICCT. Available at: https://www.icct.nl/publication/losing-plot-narrative-counter-narrative-and-violent-extremism.
12. Heath-Kelly, C. (2012). Counter-terrorism and the counterfactual: Producing the 'radicalisation' discourse and the UK PREVENT strategy. *The British Journal of Politics and International Relations, 15*(3), 394–415.
13. Lueg, K., & Lueg, R. (2021). Deconstructing corporate sustainability narratives: A taxonomy for critical assessment of integrated reporting types. *Corporate Social Responsibility and Environmental Management*. https://doi.org/10.1002/csr.2152
14. Mcfate, S. (2019). *The new rules of war: Victory in the age of durable disorder*. William Morrow, An Imprint Of Harpercollins Publishers.
15. Wilson, L. (2021). *Gone to Waste: The 'CVE' Industry After 9/11*. [online] New Lines Magazine. Available at: https://newlinesmag.com/uncategorized/understanding-the-lure-of-islamism-is-more-complex-than-the-experts-would-have-you-believe/ [Accessed 22 Mar. 2024].
16. Zamir, M. (2022). *Reconceptualizing Security Challenges*. Springer eBooks, pp.275–293. https://doi.org/10.1007/978-3-319-91875-4_42

The manufacturer's authorised representative in the EU is Springer
Nature Customer Service Centre GmbH, Europaplatz 3, 69115 Heidelberg,
Germany. If you have any concerns regarding our products, please
contact ProductSafety@springernature.com

Printed and bound by CPI Group (UK) Ltd, Croydon, CR0 4YY
01/10/2026
02225980-0016